RASPBERRY Pi

Haynes

**A practical guide to the
revolutionary small computer**

Dedicated to Karl and Corinne, with apologies for ICT.

First published in March 2013

A catalogue record for this book is available from the British Library

ISBN 978 0 85733 295 0

Published by Haynes Publishing,
Sparkford, Yeovil, Somerset BA22 7JJ, UK

Tel: 01963 442030 Fax: 01963 440001
Int. tel: +44 1963 442030 Int. fax: +44 1963 440001
E-mail: sales@haynes.co.uk
Website: www.haynes.co.uk

Haynes North America, Inc.,
861 Lawrence Drive, Newbury Park, California 91320, USA

Design and layout: James Robertson

Printed in the USA by Odcombe Press LP,
1299 Bridgestone Parkway, La Vergne, TN 37086

RASPBERRY Pi

A practical guide to the revolutionary small computer

Owners' Workshop Manual

Gray Girling

Foreword by **Eben and Liz Upton**

CONTENTS

5 Hardware Recipes

6 Meal Plans

7 Annexes

Foreword

Seven years ago, the longest piece of veroboard it's possible to buy was sitting in the middle of our kitchen table alongside some pots of home-made marmalade. The sticky stuff wiped off, that veroboard was bristling with transistors, some of Eben's very neat hand-soldering and an Atmel chip. It could drive very simple 3D graphics on an old-style analogue TV – and not much else. That prototype (called at the time an ABC Micro, but very soon to be christened Raspberry Pi) was the first in a long line of hardware iterations that has led us to the very strange place we find ourselves in now.

Today we've found ourselves with a million-unit computer business that suddenly blinked into existence when we weren't looking (bringing down all our suppliers' websites in the process on launch day and leaving us running so hard to keep up with the year's demand that it's been hard to find time to breathe), a completely new schools curriculum to help construct, and the piece of hardware you're holding in your hand today: a fully-fledged Linux PC, small enough to fit in your pocket, and cheap enough to be bought with your pocket money. Unlike that 2006 kitchen-table model, today's Raspberry Pi can play and record Blu-ray quality HD video on any TV, HDMI or analogue; it can run any Linux programming language you want to throw at it. It can interface with the real world: people are using their Pis to control heat and pressure in breweries and bakeries, to make their mobile phones open their garage doors, to send pictures back from hydrogen balloons in near space, to build car computers, to monitor their babies and to build single button audio-book devices for their grandparents. If you're feeling lazy, you can just use it to watch TV. Or you can use it to learn how to program.

We launched Raspberry Pi just under a year ago (at the time of writing). Our stated aim has always been to use it to encourage kids to learn to program, and to put the tools to do that in the way of any child who might have an aptitude, in a way that is cheap and accessible. Unlike people like us, in our middle-thirties, the kids we meet now don't have that easy access to programming that we had in the 80s, with our BBC Micros and ZX Spectrums. The machine-for-programming market got eaten from both ends. At the top it was cannibalised by the family PC, which has a central role in the family; what with Mum and Dad's work, banking and family internet use. Very few parents are prepared to let their kids take the bonnet off the household PC and tinker with what's underneath. And from the bottom, many of those 80s machines were replaced with single-purpose gaming devices. Games consoles are explicitly manufactured to be unprogrammable; they're there to get you to buy content to put on them, not to make your own for free. They're great at what they do (there are two under our own TV), but for so many families they're the only thing that a kid will recognise as a computer in the house. We wanted to make a tool that, unlike smart phones or tablets, emphasised creating rather than consuming. We wanted to make something that'd lure kids in with a few games and great multimedia capability for a very small amount of money. And we wanted to dangle the potential to make things themselves under their noses for as long as they used it.

Of course, it's not just kids who are buying the Pi (although we reckon about 20% of the units sold so far have ended up in the hands of children). We never, ever imagined that the response would be so large, so enthusiastic, or so wonderfully eccentric: in a world where someone takes the time to build a piece of furniture comprising a gorgeous poplar stand, six beetroots and a Raspberry Pi that makes beautiful music every time someone fondles one of the beets, you quickly become used to the idea that your ability to keep tabs on the shape of things has long ago passed you by. Our original goal is still firmly in our sights: this year the Raspberry Pi Foundation is working with a large number of other organisations to work on a replacement for that disastrous ICT curriculum (now abolished) that turned out a whole generation of kids whose entire interaction with computing resulted in a few spreadsheets and a couple of PowerPoint presentations. We're making real headway into showing kids that computing's not about maths: it's about creating things, be they games, parent-detecting sensors or fortune-telling boxes. (Or musical root vegetables.) We're talking to a large and growing number of teachers and teaching bodies; we're dealing with after-school and parents' groups.

The success of Raspberry Pi so far has largely been down to the community that's built up around the device. There's something approaching a post every single minute to our forums at www.raspberrypi.org, where raw beginners, old-timey electronics nerds and everybody in between seems to have something to say: we guarantee that if there's something you can't work out, someone there will have shown you what to do within the day. These are the people who run enterprises like the worldwide network of Raspberry Jams – meetups for new and more seasoned users – or like the MagPi magazine, which is written and produced by a small team of volunteers for no pay, out of simple enthusiasm and a desire to help. There are just a small number of us at the Raspberry Pi Foundation, and we're more grateful than you can imagine to all the people who have helped us out in this first year by doing something cool with a Pi and telling people about it online it, by holding Raspberry Pi meetups in schools and cafés, by writing documentation, asking thoughtful questions, or by giving one of these little computers to a child who might just be interested.

Ultimately, it's about access to tools. Given a small amount of the right knowledge (you're holding a book full of the stuff), you can build anything.

Eben and Liz Upton
The Raspberry Pi Foundation
January 2013

INTRODUCTION
History

Back in the hazy depths of history there was a time when computers filled whole rooms in university departments, when the only way to talk to them was to make holes in paper cards, and when the general public thought they were something to do with huge spinning disks of magnetic tape. It was in those Stone Age times that I took my first steps towards nerdhood. The Internet was an American military research project, email was very cool, and a phone was something Bakelite tethered to your hall by a bit of cloth-covered wire.

For a growing geek there's probably never been a more wonderful time. Every couple of years since then something exciting has doubled in speed or size or density. This isn't just a figure of speech – it's a rule of thumb which has been true for over 50 years. Called Moore's Law, it was originally coined in 1965 by Gordon Moore, one of the founders of Intel, who observed that the number of transistors on integrated circuits (chips) was doubling every two years. In 1978 it was accurate enough for people to muse that, around about now, we'd be able to keep whole libraries of films on home computers, and today you can still use it to make your own predictions as to what the future may have in store.

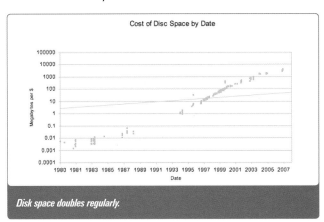

Disk space doubles regularly.

The thing that hasn't been predictable is what playthings technology will be used for, who'll get to make money from them and what difference it will make to society. Who would have guessed that processing data for the 1890 United States census would eventually give the world IBM, that organizing deliveries to Lyons tea shops would have lead indirectly to the UK computing mainframe business, or that a British physicist wanting to share information at CERN would give us the Web?

It isn't simply that great contributions come from unexpected places – great contributions famously often come from tiny initiatives. Computing is full of stories of small beginnings turning into huge successes, and the one or two people at the centre of those beginnings just needed (a) the right technical background, and (b) a dose of serendipity – coming up with the right thing at the right time. While a Raspberry Pi and this book won't give you luck it just might help with a bit of the technical background.

The ARM Age

If you have a long memory you might recall that the most common sort of computer was once made by IBM. There was a long dynasty of huge ('mainframe') computers which all had one thing in common: they all understood the same kind of computer instruction.

With the rise of personal computers a new dynasty was established based on a new set of computer instructions invented by Intel. If you've ever used the Windows operating system you've most likely been running on a computer that understands these instructions. As with IBM, the instruction set gradually evolved, adding extra instructions with each new processor released. Nowadays the newest Intel processors can still understand the instructions meant for the earliest. The whole gamut of instructions is referred to, rather grandly, as the Intel x86 architecture.

Today, it might surprise you to hear, the x86 architecture is by no means the most used instruction set in the world. That honour belongs to the ARM architecture. ARM processors are everywhere. You almost certainly possess several of them yourself. In 2011 there were said to have been some 15 thousand million (billion) of them produced. Since the population of the world is estimated to be around seven billion this means that there's now been at least one ARM processor for every actual arm of a living person. Computing archaeologists of the future sifting through the detritus of our age might well call our own times the 'ARM Age'.

If you want to take an active part in the ARM Age what you need is an ARM to experiment on, and luckily this is the very processor that Raspberry Pi uses.

The BBC Micro

'ARM' once stood for Acorn RISC Machine, having originally been created by an offshoot of Acorn, the company who made the BBC Micro. Since then (and the sad demise of Acorn) the acronym has become Advanced RISC Machine.

Britain is full of oldies who grew up using the BBC Micro who fondly remember it kindling their interest

The BBC Microcomputer.

in all kinds of things computing. It was commissioned by the BBC as a PC to support their computer literacy project; they wanted something they could use in their TV series *The Computer Programme* during the 1980s.

I first came across the BBC Micro when one of my colleagues, Ram Bannerjee, in the computer laboratory at Cambridge University helped construct the prototype circuit boards. Acorn won the right to call their machine the 'BBC Micro' only after meeting some very tight deadlines in a competitive tender. Ram used the fruits of a research project that helped wire-wrap components together really quickly. I was lucky enough to be involved in some of the pre-production testing, and later joined Acorn to work on it.

If you read our companion Haynes volume *The Video Gaming Manual* you'll find a comprehensive history of retro games platforms. You won't find the BBC Micro there, though – you'll find the cut-down Electron instead. You could run games very well on the BBC Micro (and people did) but, true to its educational roots, it invited users to make their own. When you turned the computer on the first thing you saw (very quickly, by today's standards) was a prompt, asking for your first line of a program. If you've always used a computer whose first request is that you should select a program you want to run, you won't have noticed the sleight of hand that you've been

```
BBC Computer 32K
Acorn DFS
BASIC
>
>10 PRINT "HELLO WORLD"
>
>RUN
HELLO WORLD
>
>
```

Turn it on and program.

subject to. I remember how lost I felt when I used my first Windows operating system (Windows 95) and found no immediate way to control what it was to do for me; no prompt; no programming language – a computer that belonged to some other programmer, not to me.

The dark years

For years since the '90s the computers in people's homes have invited them primarily to consume the programs that are provided for them. When people need something done, even when it would be easy to achieve it by combining or tweaking the programs they already have, it's now far more common to either search the Web for a new program that does precisely what's needed, or to just give up.

Missing the opportunity to make such a 'tweak' yourself is more or less insignificant on a single occasion, but computers aren't about to disappear, and a lifetime of not doing simple things (that could inevitably have developed into more complex things) amounts to an entire lost education. So, not knowing where to start with such tasks forms a small but effective barrier that turns potential producers of new programs (and thence technology in general) into fodder for the software market.

To compound the message 'you are a consumer, not a producer' in children, the British education system has for many years standardized on an Information Communication and Technology (ICT) curriculum that, to many, seems to emphasise the use of de facto standard programs with little or no attention paid to the much more interesting detail of how these programs came to be produced in the first place.

The Raspberry Pi 'project'

By the time children leave school they have little idea of the basic technology behind the tool that they've effectively been told is good only for word processing, spreadsheets and databases.

After a bulge in the population of motivated and informed people in the '80s and '90s, university computing science departments are now finding an increasing number of school leavers applying who have no real experience of recreational programming.

The people behind the Raspberry Pi Foundation (a UK charity) see the provision of a truly personal computer as a way to try to redress this situation in the same way that the BBC Micro did some 30 years ago.

Amazing technology

The chips inside Raspberry Pi are amazing. In fact, thanks to Moore's Law all the chips made for everything are more amazing than they were last year. Over a year or two the difference adds up, and might amount to enough to make you want to throw away your old camera, laptop and television to buy another one. Over a decade or two, though, the differences verge on the boggling. What originally came in huge cabinets or perhaps rooms full of cabinets now sits in a tiny corner of a chip.

My first university (in the late '70s) had one huge CDC 7600 computer that serviced all of its terminals. It was a cutting edge design containing a number of separate computer processors and was said to be able to perform 36 million mathematical (floating point) operations in a second. At the time we were in awe of the upcoming next generation that would incorporate a 'vector' unit that could replicate the same operation on a series of data values many, many times just by obeying a single instruction.

Because Raspberry Pi doesn't include 120 miles of hand-wired connections, or consume hundreds of kilowatts of power, it costs quite a bit less than the millions of pounds the CDC 7600 did. Nonetheless, it too is a multi-processor containing two extra 'VideoCore' processors which each have quite nippy vector processors in them, each capable of 24,000 million floating point operations per second and, instead of using the kind of electrical power that could have run a whole village, the chip in Raspberry Pi can be run off a mobile phone battery. Among its peers it isn't so much the tiny size or impressive computational clout that the Raspberry Pi chip is known for, but rather the very low power it needs to do what it does.

Doubtless this chip would have been put to service running a university too if it had been available in the early '80s. Today, though, its purpose is to fit neatly in a mobile phone.

The chip in Raspberry Pi

There are very few chips in Raspberry Pi, and almost everything important is contained in just the one, in the centre of the board:

Looking carefully you might be able to see that this chip has two layers, a bit like a liquorice allsort. The top contains all the computer's memory. The bottom may look small but it contains a lot

The BCM 2835 chip at the heart of Raspberry Pi.

of electronics. The features on the chip are so tightly packed that the distance between them is only 40nm. Drop a hair on its surface and it might cover 1,000 rows of electrical components.

As well as the ARM and the two VideoCores it contains everything needed to help the VideoCores deal with video and audio received from various sources and sent to several destinations.

The ARM itself is asked to execute an astonishing 700 million instructions every second. To give you an idea of how fast this is, in the same time as it takes to execute one instruction light could travel only about 43cm (which explains why chips can't be very large).

With a name like 'VideoCore' you might expect this part of the chip to be pretty good at TV and video output, and you'd be right. It's quite comfortable providing video in the highest definition of any commonly available source (such as DVDs or Blu-ray discs). Its 3D graphical performance is at least that of the previous generation of games consoles, which is an amazing feat for a chip that doesn't need any special measures to keep it cool.

Free as in...

In the years since the BBC Micro was launched and Moore's Law changed the face of computer hardware, a separate movement has been at work on computer software: the movement towards 'free' software.

This isn't, as you may think, simply to do with the fact that everyone likes to get things that don't cost money, because the 'free' in 'free software' isn't always supposed to mean 'free' as in 'free beer'. Sometimes it's meant to mean 'free' as in 'free speech' or 'free man': that is, software that has 'freedom' and isn't unduly encumbered by things like licensing, patents or legislation.

As computers have become more and more numerous the audience for an altruistic programmer has grown and grown. Many find the attraction of being able to help so many with their programming efforts irresistible, and this has resulted in thousands of free software projects. International intellectual property legislation (such as copyright) gives authors rights over the programs they write and, unless something's done, the legal aspects of big projects with hundreds of contributors can quickly become impossible to manage. Groups such as the Free Software Foundation work very hard to provide special licences that allows free software to be developed, and fight legislation that threatens that possibility. This is important because there are many ways that a 'free' (as in beer) program can change into a paid-for program if it isn't also 'free' (as in speech).

There are now many significant free software initiatives that equal, if not exceed, the abilities of commercial equivalents. More importantly, though, it's becoming more and more accepted that the very open methods that have to be used in large 'free' software projects have implicit advantages over the closed development that's necessary in commercial projects. The large number of developers and their mutual interrogation, along with the sheer number of people available to spot errors in the programs, results in levels of correctness that are difficult to achieve commercially.

The Linux operating system and the Python programming language at the heart of Raspberry Pi are examples of such free or 'open' software (in fact virtually all Raspberry Pi software is), and to some extent one of the aims of the Raspberry Pi project is to promote the use of open software.

Large programs, like Linux or Python, are complex, and using them will involve developing skills that represent both a cost to you (since you have to put a lot of time into acquiring them) and a benefit (since you'll be able to use them in the future). The benefit is larger the more you can use your skills in the future, but smaller if the program stops being available or starts being too expensive for you to use. One of the great advantages of open software is that both of these risks are greatly reduced. There would be little point in learning to use Python for everyday small tasks – for example, if it started to cost more than those little tasks justified, or if it might suddenly be taken off the market. Choosing to learn the use of 'free' programs is a safer investment than choosing commercial equivalents.

Book layout

Anyone remotely interested in their new toy will first and foremost want to get it working, so, right after this introduction, the first chapter is all about putting everything together and getting started.

This book, though, isn't just a manual for the already-capable to make use of the Raspberry Pi single-board computer; it's a collection of parts that will help to remove the barriers between any motivated person and his or her desire to fiddle with it in as many ways as possible.

Such a broad aim is, of course, impossible to fulfil completely in one book, so many of the sections are 'just enough' to provide a start, which you can take further using more comprehensive detail available elsewhere.

Many of these barriers aren't directly to do with Raspberry Pi itself – they're a simple question of learning about various bits of technology; so, to use a cooking analogy appropriate for a raspberry pie, the next chapter will be cooking lessons: introductions to programming languages and operating systems. If you already have a background in such things you could omit them.

A raspberry pie isn't a meal in itself. It's part of some greater meal plan. Raspberry Pi could have been built to be the centrepiece of many a project. Yours might bring together many different recipes to make something of your own. To that end the next chapter of the book concentrates on providing little recipes that give you enough information to gather a whole pantry full of useful bits and pieces.

The following chapter concentrates on more established software recipes that'll help if your project needs some way to provide an interface to a remote user, do things regularly or provide various network services.

The last chapter looks at hardware recipes for using the many plugs and sockets on the board, including a number that are hidden among the hedgehog of pins in the corner of the board. A little bit of software is explained for most so that you might be able to use them in programs of your own. The fraught issue of running off batteries is also discussed.

Finally the book provides some simple examples of meal plans that you might be interested in. Some, like the game of 'snake' and the toy that provides twitter alerts, are a start that you can build upon. Others like the fully working media centre are ready-made projects that just need to be used.

And finally, you can find the program code for many of the examples in the book at **www.haynes.co.uk/rpiresource**.

02.

A quick bite

A QUICK BITE
A nibble of your Pi

L et's put your new Raspberry Pi together for the first time and make it do something.
There are three main parts to getting things operational:

- Getting the software Raspberry Pi will need on to a Secure Digital (SD) card.
- Assembling the Secure Digital (High Capacity) card.
- Running for the first time.

Below we cover each of these one by one. Inevitably you may have problems that weren't foreseen here. When that happens your favourite Web search engine is your friend. One place to try early is the Raspberry Pi Forum site at:

A Secure Digital (High Capacity), or SDHC, card

```
http://www.raspberrypi.org/
```

Software assembly

If your Raspberry Pi came with its own SD card you should skip this section and go on to the next one.
There are several versions of Linux available for Raspberry Pi. Those that are officially recommended are available at:

```
http://www.raspberrypi.org/downloads/
```

Your first task is to download one of these on to a computer; then you can write it to an SD card.
Naturally you'll need a blank SD card for this stage. SD cards come in a variety of sizes. The software for Raspberry Pi will use less than 2GB, but the SD card is also used to store your own files, so the larger the SD card the better. A 4GB card should be considered a minimum size.

Your download computer

You can use either a Windows or a Linux machine to create the SD card. The main features that it should have are:

- Internet access.
- Either a built-in SD card writer (many laptops do) or a separate SD card writer (there are many available that plug into a USB port on your computer).
- Enough spare disk space to hold the SD card image (at least 2GBytes free should be enough – this is very unlikely to be a problem unless you have a very old computer).

Downloading the SD card image

Choose the system you need from those listed at:

```
http://www.raspberrypi.org/downloads
```

This book assumes that you'll download a 'wheezy' version of Debian Linux. For example, one like that indicated here:

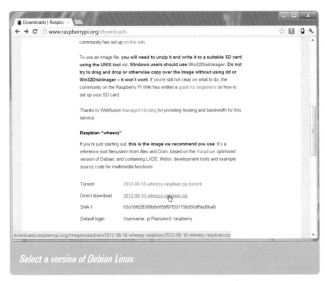

Select a version of Debian Linux.

This will lead you to a page where you can make the download. The file will be a large '.zip' file, so it may take some time for the download to complete.
Make a note of where your browser has put your download (often it'll be in your 'Downloads' directory, both on Windows and on Linux) – you'll need to unpack it next.

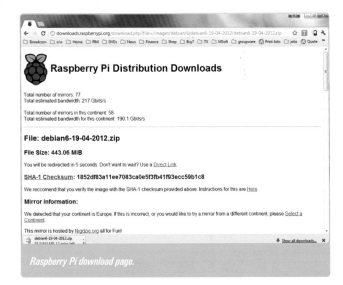

Raspberry Pi download page.

Writing SD cards is dangerous

No matter which operating system you use, a program capable of writing an SD card image is also capable of replacing the entire contents of a hard disk.

WARNING

It's very important that you ascertain the SD device name correctly. If you select one of the hard disks by mistake the damage you do will be impossible to repair.

Writing the SD card from Windows

Open a folder display in the directory where your download was placed and unpack the '.zip' file. Most versions of Windows will provide an 'Extract' or 'Extract All...' option to do this for you when you right-click on the file name.

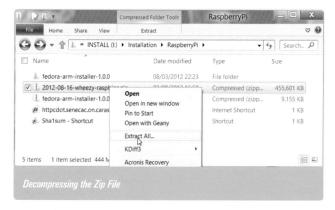

Decompressing the Zip File

No 'Extract All...'?

If you have an older version of Windows or do not see another way to unzip the folder look for the free program called '7-Zip' on the Web, download it and install it. It will have a '7-Zip' sub-menu from which you can select a number of options to unzip the file.

The recommended program to use to write the disk image file to an SD card is called 'Win32DiskImager' available, if you haven't installed it already, at

```
https://launchpad.net/win32-image-
    writer/+download
```

You will need the file called 'win32diskimager-binary.zip', which you need to download and unzip in the same way that you did the SD card image. When this file is unzipped a number of files are generated including a program called 'Win32 Disk Imager'.

Run the disk imager program by clicking on it. As well as knowing the location of the disk image file you will also have to know the name of the device that holds your blank SD card.

Run Win32 Disk Imager.

It should display a dialogue like this one asking for the image file.

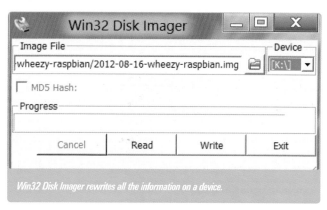

Win32 Disk Imager rewrites all the information on a device.

Click on the icon of a folder and locate the Raspberry Pi image file you downloaded. Then select the 'device' where your blank SD card has been placed (noting the warning above). Win32 Disk Imager rewrites all the information on a device.

Win32 Disk Imager provides some help by listing only 'removable' devices that could be SD card writers. If you don't have too complex a computer it might list only one device. You may be able to determine the correct name of the SD writer by observing the name of the folder that opens when the SD card is inserted into the reader. Alternatively you might be able to spot a change in the 'Computer' or 'My Computer' display as you either insert and remove the SD card from the writer, or insert and remove the writer from the computer (eg if it's a USB device).

Once you've identified the correct device and the SD card image click on the 'write' button and wait for it to complete (this could take several minutes).

Win32 Disk writing an SD card.

Writing the SD card from Linux

You can locate your SD card by seeing which devices are identified as containing partitions before and after the SD card is introduced into the computer. This command will print all the partitions that Linux has identified:

```
cat /proc/partitions
```

Use it once before inserting an SD card and once afterwards. The lines that are added will identify the device that you've plugged the SD card into. For example, if these were the partitions before the SD was inserted:

```
% cat /proc/partitions
major minor    #blocks  name
    8     0  312571224  sda
    8     1   15360000  sda1
    8     2   26291544  sda2
    8     3          1  sda3
    8     5    9764864  sda5
    8     6  261151744  sda6
%
```

and these were those afterwards:

```
% cat /proc/partitions
major minor    #blocks  name
    8     0  312571224  sda
    8     1   15360000  sda1
    8     2   26291544  sda2
    8     3          1  sda3
    8     5    9764864  sda5
    8     6  261151744  sda6
    8    16    1931264  sdb
    8    17    1930680  sdb1
%
```

we can see that devices sdb (the whole SD card) and sdb1 (partition 1 of the SD card) are new. These names correspond to the devices /dev/sdb and /dev/sdb1. The device we want to write the SD card image to is /dev/sdb. (Note the warning above.)

Some Linux installations will automatically mount SD cards that are inserted into the system, but the next stage needs the device to be unmounted. Mounted devices are normally visible in the list in the file /etc/mtab. If the SD card has been mounted, unmount each of its mounted partitions (such as /dev/sdb1) using:

```
sudo umount /dev/sdb1
```

Next use the 'dd' command to write the image to the device. If the image file was called '2012-08-16-wheezy-raspbian.img' and the SD writer had been found in device '/dev/sdb' (not '/dev/sdb1') the command would be

```
sudo dd bs=1M if= 2012-08-16-wheezy-
    raspbian.img of=/dev/sdb
```

This can take several minutes. The command doesn't generate any output during this time.

Hardware assembly

If you don't have all the equipment you need for Raspberry Pi you can usually buy what you need quite cheaply (with the exception of a display), or if you already have a computer you may be able to try things out simply by borrowing equipment from that.

You may be used to buying equipment for your Windows computer and finding that almost everything on sale will function correctly when you install it (or you may not). Sadly this level of success isn't to be expected when buying for Raspberry Pi. To improve your chance of buying equipment that'll work properly try looking at this Web page, which lists the things that have and haven't been used successfully in the past:

```
http://elinux.org/RPi_VerifiedPeripherals
```

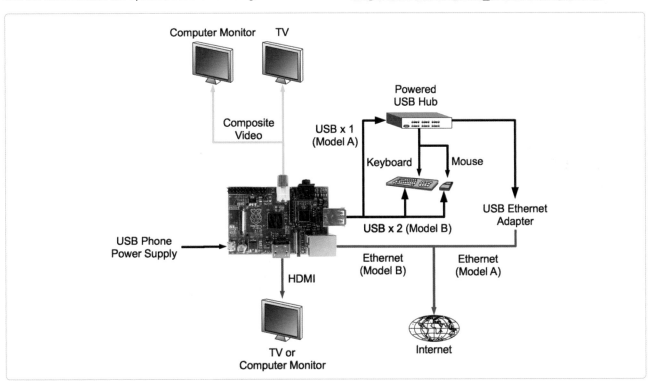

We'll talk later on about the many options for connectivity that Raspberry Pi supports – here we concentrate only on what you need to get going.

Apart from a Raspberry Pi single-board computer you'll need the following:

- A USB power supply.
- The SD card containing Raspberry Pi software.
- A display (eg a TV).
- A USB keyboard.

The following would also be useful, especially during first setup, but aren't strictly necessary:

- A USB mouse.
- An Ethernet connection to the Internet.

It's possible to use a USB WiFi adapter to connect to the Internet using a USB socket instead of an Ethernet adapter, but it must be selected very carefully – many of them can't be set up properly without first having an Internet connection! A later section deals with the use of WiFi adapters.

Raspberry Pi Model A

There are two types of Raspberry Pi available: a cheaper one (Model A) with only one USB socket and no Ethernet; and another (Model B) with two USB sockets and an Ethernet port. Model B contains twice the amount of memory as Model A too, unless you have an early version of Model B (which has 256MB, like Model A).

There are many things that you can do with no network connection and only a keyboard connected, but with no mouse and no network some things are difficult (such as updating the software and using the graphical desktop).

If you do want to update Raspberry Pi's software during the setup process you'll need to borrow or buy some extra equipment:

- A powered USB 2.0 Hub which will give you enough USB ports to plug in a USB keyboard and a network adapter (and a mouse if you want one).
- A USB Ethernet adapter which will enable Raspberry Pi to connect to the Internet.

A powered USB hub not only supports a number of extra USB sockets (typically four) but also provides them with power from its own power supply. Always remember to have the USB hub's power supply attached and turned on when you're using Raspberry Pi. Because the Raspberry Pi is itself powered from a USB plug it's possible to minimize the number of connections by using one of a powered hub's sockets to provide the power that Raspberry Pi needs. This will require the powered hub to provide up to 700mA, however, so if you're planning to buy a new one you should check that this is possible.

The USB port on a Model A can supply only a very small current and this may be insufficient even for some complicated keyboards. If Raspberry Pi fails to work properly with your keyboard or behaves erratically try another keyboard or use a powered USB Hub.

SD card

The SD card that you've created above is inserted in a socket underneath your Raspberry Pi.

SD cards differ in both the amount of information they can hold and the speed that they allow information to be used. You'll probably see an

The SD card slot with an SD card inserted.

indication of the size and speed of your SD card on its label. Today it's unusual to find cards that can hold less than 2GB (2,000 million bytes), and other sizes each twice the last are commonly available (4GB, 8GB, 16GB and 32GB). Cards of all of these sizes can be used on Raspberry Pi.

The speed of an SD card is often indicated by its 'class', shown as a small number inside a capital 'C'. This number shows you how many million bytes can be read from the card in a second (MB/s). Class 2 and Class 4 are common, and higher classes aren't rare. Some Class 6 cards have worked poorly on Raspberry Pis so it may be sensible to use only the slower Class 2 and 4 cards.

Power supply

USB power supplies are sold as chargers for a variety of phones, cameras and other devices. In general they're interchangeable, all delivering 5V (but you should check – Raspberry Pi needs exactly 5V to the nearest quarter of a volt). The power the supply can generate, usually rated in milliamps (mA), must be greater than 700mA, but at least 1,000mA is recommended.

In principle it's possible to power Raspberry Pi from a powered USB hub instead of a power supply, or even a USB port on a computer. Unless these are explicitly designated for supplying power, however, they aren't really suitable and may not supply the quantity of power Raspberry Pi requires. A typical computer USB port can provide only up to 500mA, although there are exceptions.

The cable on the power supply should support a 'micro B'-style USB plug.

Micro A – bad.   *Micro B – good.*   *Mini B – bad.*

If you have a power supply with a separate body and cable be very cautious with the cable you buy. Some cheap cables are very poor at delivering power (because the wires they use have a high resistance).

Batteries can be used to provide power to Raspberry Pi, but only with some degree of care. We discuss running from batteries in a later section.

HDMI display

As long as it supports High-Definition Multimedia Interface (HDMI) you can connect Raspberry Pi to either a television or a computer monitor. Modern digital televisions usually support one or more of these. The HDMI socket on Raspberry Pi is here:

The HDMI socket.

An HDMI socket on your display might look like this:

Example HDMI sockets on a television.

An HDMI cable is one with a plug like this at both ends:

Your computer's monitor may have a DVI connection instead of an HDMI one. If so you may be able to use this via an HDMI-DVI lead or adapter.

An HDMI plug suitable for Raspberry Pi.

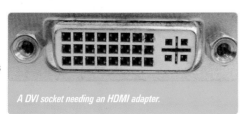
A DVI socket needing an HDMI adapter.

Composite video display

You'll usually get a much better display if you can use HDMI, but if you have an older TV you may not have one. In that case you may find that your television (or sometimes, in older models, your computer monitor) has a composite video connection.

A composite video lead is usually one with an RCA (or 'phono') connector plug like this at each end:

Your display is likely to have a socket that's also coloured yellow (if it supports

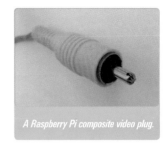
A Raspberry Pi composite video plug.

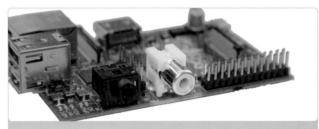
The composite video socket.

composite video directly). Many televisions, however, support composite video through the much larger SCART connectors. This is what a SCART plug looks like:

This is actually the SCART plug side of a SCART-to-phono adapter which provides two sound connections (one for your left ear and one for your right ear) and a video connection. This is the other side:

Unlike an HDMI lead a composite video lead doesn't carry sound information. Sound is still available from the Raspberry Pi using the 3.5mm jack plug and this could be connected to the other two phono sockets above with an appropriate lead.

A television SCART plug.

Phono connectors on the back of a SCART-to-phono adapter.

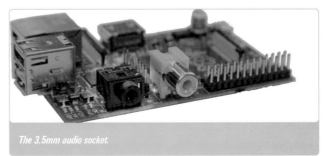
The 3.5mm audio socket.

No display

Believe it or not you can use Raspberry Pi without any display at all (which is much cheaper but won't allow you to show off its video prowess). There are several ways to do this:

■ Use the serial port (the UART) to connect to another computer (how the UART can be used like this is described in its own section).

■ Use the network to make an SSH terminal connection (setting this up initially is described below, and there's a section about it later).

■ Use the network to provide a graphical interface using X, RDP or VNC (each are described in their own sections too).

For setting Raspberry Pi up for the first time without a display, though, you'll need to use the serial port (we describe how to make one of these later).

A Raspberry Pi serial connection.

USB keyboard

If you have a USB keyboard it'll have a plug like this:

Most keyboards like this will work with Raspberry Pi. Beware the keyboards that incorporate USB ports of their own, touchpads, or other gadgets, which may require too much power for a Model A to drive. The USB sockets on the Raspberry Pi are here:

Keyboards with all kinds of extra buttons and other features often require specialized drivers that sometimes are available only for the Windows operating system. This usually means either that those keyboards won't work on Raspberry Pi, or that those extra features won't be available.

USB keyboard plug.

The USB sockets on a Model B.

PS/2 keyboard

Some older keyboards have a PS/2 plug. Unfortunately these keyboards usually require a slightly greater voltage to work than the Raspberry Pi provides, so even with an adapter cable they may not work. If they are connected via a powered hub, however, they may work.

A PS/2 to USB keyboard adapter cable.

No keyboard

If you have a display you can still use Raspberry Pi without a directly connected keyboard by using the one connected to another computer that you use via the serial port. The way this 'UART' is set up is described in a later section.

Mouse

Mice come with the same range of connections and problems as keyboards. A PS/2 mouse can be converted to USB via a PS/2 to USB connector cable and will probably also need to be connected via a powered USB hub.

It's easy to set up Raspberry Pi for its first run without a mouse, but not having one might limit what you can do with it in the future. If you plan always to use your Raspberry Pi remotely you won't need a directly connected mouse.

Wireless mouse and keyboard

There are a number of wireless mouse and keyboard kits that come with a single USB device to plug into Raspberry Pi (sometimes called a 'dongle'). Many of these will work correctly on the Model B and may work on the Model A if they aren't too demanding in terms of power.

It's possible to use a Bluetooth mouse and keyboard with Raspberry Pi (as described in a later section) but this isn't recommended for the first time the system is used.

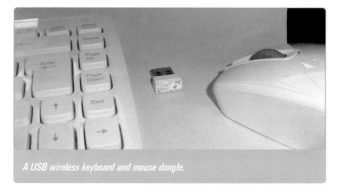

A USB wireless keyboard and mouse dongle.

The first run

Once you have all the equipment, plug in your keyboard and display. Also plug in your mouse and Ethernet plug if you have them. You should turn on your television or monitor and select its source of video if necessary. Finally, ensure your USB power supply is plugged into the mains and turned on, then attach the USB power cable to Raspberry Pi. Raspberry Pi has no on/off switch: you turn it on and off by attaching and detaching the power cable.

If all is well the power LED will come on (marked PWR on the board). If it doesn't, there's probably something wrong with your power supply.

Raspberry Pi should do a quick display test (you should see a multi-coloured rectangle flash) then read the operating system

Raspberry Pi (version one) LEDs.

from the SD card and start Linux. As soon as this is done the top LED, marked 'OK' ('ACT' on new boards), starts to flash whenever the SD card is being read or written. If you see no flashing and nothing appears to be happening, there may be a problem with your SD card.

If there was a fair amount of flashing but there's nothing on your display, there may be something wrong with your display. HDMI connections are much more complicated than composite video ones and they may be able to be reconfigured. It's possible to update your SD card to change the way an HDMI connection is set up. This is discussed in the annexe about configuration.

As soon as the operating system is active it will put a friendly Raspberry Pi symbol at the top of the screen and start to write all kinds of information about what it finds, and then, once only, it'll usually run the special program 'raspi-config' that will give you a menu of things that you can do to set up your system. This is a good time to perform a lot of these actions, and you won't automatically be presented with the option again. Nonetheless, you can simply skip it all by pressing the right arrow key on your keyboard until 'Finish' is outlined and then pressing return. If you do, it's still easy to perform the actions later by running the command

```
sudo raspi-config
```

The options available are described later in this section.

If you're using an SD card larger than 2GB (you probably are) it would be sensible to choose the 'expand_rootfs' option. This will allow Linux to use more than just the first 2GB of your SD card.

You might also think about using the 'overscan' option if the display screen either seems underused, with a big border, or overused with text missing from the edges.

After you've completed any initial configuration you want to do, you're given a prompt to supply your name and password. In the standard build the name 'pi' and password 'raspberry' will let you use the system. You should see something like:

```
raspberrypi login: pi
Password:
Last login: Tue Sep  4 21:30:29 BST 2012 on
ttyAMA0
Linux raspberrypi 3.2.27+ #102 PREEMPT Sat
Sep 1 01:00:50 BST 2012 armv6l


The programs included with the Debian GNU/
Linux system are free software; the exact
distribution terms for each program are
described in the individual files in /usr/
share/doc/*/copyright.


Debian GNU/Linux comes with ABSOLUTELY NO
WARRANTY, to the extent
permitted by applicable law.

Type 'startx' to launch a graphical session

pi@raspberrypi:~$
```

The last line is a 'prompt'. You're expected to type a command and then the 'Enter' key on the keyboard (sometimes indicated only by the symbol ↵). The commands available are discussed later in the section about Linux.

Using the desktop

If your system has a mouse and a display you may want to get a taste of Raspberry Pi's graphical desktop.

If your computer display looks something like this:

Textual log-on screen.

you are using the 'command line interface', which will allow you to type textual commands to Raspberry Pi. To use the alternative 'desktop' graphical user interface, log on, if necessary, and type

```
startx
```

which should start a window manager, in this case 'LXDE':

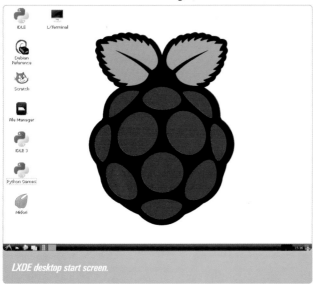

LXDE desktop start screen.

Using a command-line window

Having started a Graphical User Interface (GUI) on Raspberry Pi it's still possible to type commands into a Command-Line Interface. To do this you should click on the icon at the bottom left of the screen (a menu should appear) and select 'Accessories' followed by 'LXTerminal'.

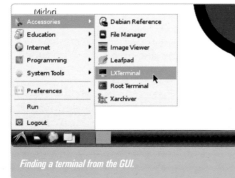

Finding a terminal from the GUI.

A window looking like this ought to appear:

A command-line interface.

All the same commands that could be used before starting the desktop can be used in this window. Running them from the desktop, however, has the advantage that these programs can also use the desktop's graphical environment.

Your boot-time user interface preference

As mentioned above, the standard Debian Raspberry Pi release contains a utility called 'raspi-config'. It can be used to set your preference for using a Graphical User Interface so that your Raspberry Pi will always boot to your choice of either a command-line or a graphical interface.

Configuring Raspberry Pi

The 'raspi-config' program which runs when Raspberry Pi boots on the first occasion isn't a standard part of Linux, and if you don't have the standard distribution this command may not be available. If you do have it, however, it's run using the command:

```
sudo raspi-config
```

The menu is operated through the arrow keys on the keyboard. Up and down select different menu entries and right and left go between the main menu and the options listed at the bottom of the screen. An entry is then acted upon by pressing the 'Enter' key.

Expand rootfs

When first written the SD card image is normally smaller than the SD card and so much of it remains unused. Selecting this option will allow the unused parts of the SD card to be used. The new SD card arrangement won't be used until Raspberry Pi is next booted though.

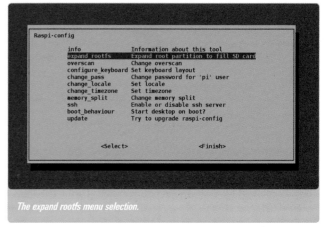

The expand rootfs menu selection.

Overscan

Some television pictures are a little rough around the edges and to make them look better a television will often ignore the outer parts of the picture. To compensate for this Raspberry Pi puts a blank area around the outside of the pictures it produces in a process called 'overscan'. It's possible that the overscan isn't necessary, so this option allows you to enable or disable it.

If the overscan is too little some useful part of the picture may still be missing from the edges; and if it's too much there

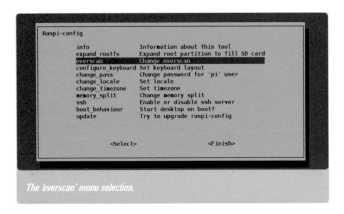

The 'overscan' menu selection.

symbols are most appropriate. Other items derived from the system's locale include the currency symbol, how numbers are to be printed, the preferred units of measurement, the typical sizes of paper and so on. The name of the locale set in the reference Raspberry Pi software is 'en_GB.UTF-8', which will select the kind of English spoken in Great Britain and the character symbols that can be displayed in the 'UTF-8' system. This menu item will allow this to be changed to an alternative.

may be a wide unused border surrounding the useful area. We cover a more advanced way to adjust these borders in a later section about booting and configuration.

Configure keyboard
When Raspberry Pi starts it assumes that it has a standard international 105-key keyboard. If you find that keys such as '@', '#', '£' or 'I' don't produce what you'd expect you have a different kind of keyboard. This menu item will allow you to change it. Expect a short delay while Linux draws up the very long list of possible keyboards for you.

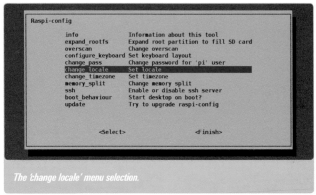

The 'change locale' menu selection.

The same function can be achieved in Debian Linux using

    sudo dpkg-reconfigure locales

Change timezone
Raspberry Pi lacks the hardware that most computer systems have to remember what the time is while it is turned off. Instead it is reliant on 'time servers' located all over the world on the Internet. Therefore Raspberry Pi won't know the time until it's plugged into the Internet and, once it is, the system must know which timezone it's in to display the time correctly.

This menu item will set the timezone by allowing you to select a nearby city, first selecting the correct continent from which to choose it.

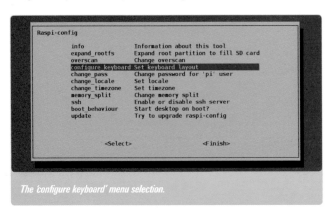

The 'configure keyboard' menu selection.

If you don't have raspi-config the same function can be achieved in Debian Linux using

    sudo dpkg-reconfigure keyboard-configuration

Change password
The password for user 'pi' appears in this book so it is safe to assume it's not terribly secret. If you have any interest in security, it is best to change it to something else. Later in the book we explain changes that, while useful, could allow other users to spend your money if they can log on as you.

The same function can be achieved in Debian Linux using

    passwd

Change locale
In most computer systems details of its locale are used to determine many things that make where you live different from where others live. Perhaps the most important of these are which language is to be used and which character

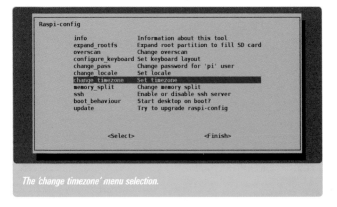

The 'change timezone' menu selection.

The same function can be achieved in Debian Linux using

    sudo dpkg-reconfigure tzdata

Memory split
The memory in Raspberry Pi is shared between the processor that runs the operating system and the Graphics Processing Unit (GPU). Depending on how much you intend to use either

part of the computer you may want the graphics part of the computer or the operating system part of the computer to have more memory. This menu item gives you a choice of a number of possible memory splits (you type in how much of your memory should be used by the GPU).

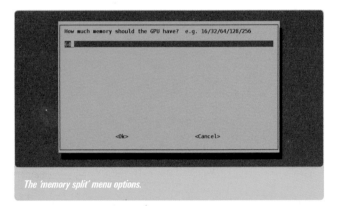

The 'memory split' menu options.

If you are going to play a lot of media (music and videos), you will want a split that favours the GPU (preferably giving it 128MB if you are going to use HD video). If you intend to use Raspberry Pi for web browsing and word processing, you will want to favour the operating system (perhaps giving the GPU just 16MB). Providing 64MB for the GPU is a good compromise. If you have a Model B with 512MB of memory you may feel it worth giving 128MB to the GPU all the time.

SSH

As mentioned above one way to operate Raspberry Pi without a keyboard (or display) is to log on remotely from another computer using the Secure SHell. You may also want this for other reasons that we cover in the section about access using SSH. This is only possible if the 'ssh server' is started whenever Raspberry Pi boots.

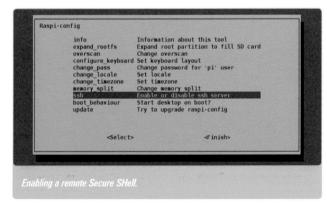

Enabling a remote Secure SHell.

By default SSH (and all other ways of accessing the system remotely) is turned off for security. This menu item enables just SSH access, which will allow others to use Raspberry Pi if they know the name and password of one of its users.

Boot behaviour

To select a preference for starting with a command-line or a graphical interface select 'boot_behaviour' (using the up and down arrows on your keyboard) and then press the 'Enter' key.

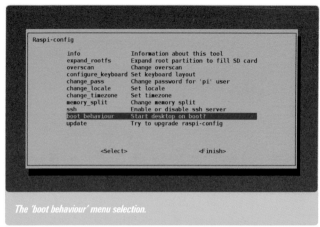

The 'boot behaviour' menu selection.

Make the appropriate choice in the next screen. You should then find that Raspberry Pi boots directly to the selected type of interface when it is next started.

If you don't have raspi-config the same function can be achieved in Debian Linux using

Boot to desktop or to command-line interface?

```
update-rc.d lightdm enable 2
```

to use the graphical display manager on start-up, and

```
update-rc.d lightdm disable 2
```

not to use it.

Update

Because Debian programs are being updated continuously the software that was placed on the SD card is almost certainly out of date. Selecting this menu item will fetch information

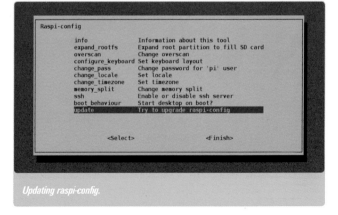

Updating raspi-config.

from the Internet that tells us which software needs updating. It will also load a newer version of raspi-config if one has been created. If you don't have raspi-config the update function can be achieved in Debian Linux using

```
sudo apt-get update
sudo apt-get install raspi-config
```

Replacement software can then be fetched and installed using

```
sudo apt-get upgrade
```

Advanced configuration

The SD card is normally split into two parts: a small one, the 'boot partition', visible on Windows computers (if you use the SD card on Windows); and the 'root partition', used only for Linux. In Linux the boot partition normally appears as the directory '/boot'.

Configuration files on the boot partition provide a great many options for the way that Raspberry Pi is set up before Linux starts, including details that you may need to make 'difficult' HDMI displays work.

Further configuration options deal with over-clocking (making Raspberry Pi go faster than it ought to), GPIO and UART setup, configuration of the SD controller, and the licensing of codecs to allow playback of certain kinds of video. The many options are detailed in an annexe along with information to help you recover from various kinds of calamity.

Other plugs and sockets

When putting everything together for the first time you don't need to make use of all of Raspberry Pi's hardware. There are a number of extra plugs and sockets you can use, with the full picture illustrated below:

Your Raspberry Pi can interact with your own electronics by using pins on its General Purpose Input and Output (GPIO) connection. Although the GPIO connector has 26 pins not all of these are used. We'll discuss exactly how the pins can be used later, in a section about GPIO.

Small groups of these pins can be dedicated to serve as particular kinds of hardware interface (in which case they aren't available for general purpose input and output).

The GPIO pins

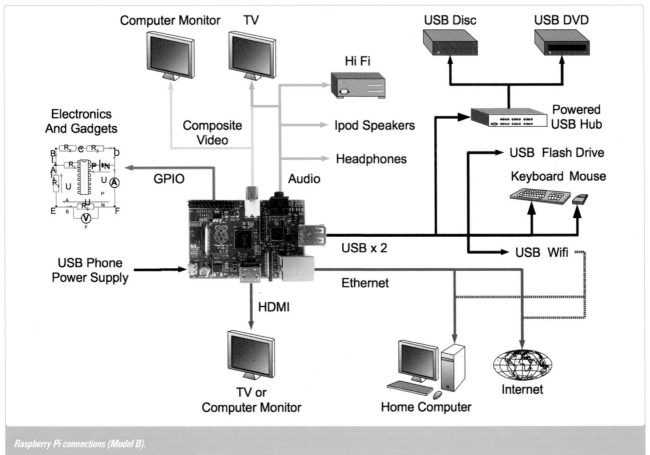

Raspberry Pi connections (Model B).

SPI and I2C

Two of the groups of pins with special functions allow your Raspberry Pi to interact with various gadgets that use the very simple Serial Peripheral Interface (SPI) and another group with the even simpler Inter-Integrated Circuit (I2C) interface. These can be used separately or together, although normally the pins they use aren't assigned to these specialized uses. There are two separate sections explaining how these interfaces can be used.

LCD display

Naturally the mobile phones that Raspberry Pi's main component was designed for use an LCD display and the board includes the hardware Display Serial Interface (DSI) that it uses on one of the two 15-way flat flex connectors. At the time of writing, however, ths software required to drive this interface has not been released.

The Display Serial Interface connector.

Camera

There's a 15-way 'flat-flex' connector on Raspberry Pi that provides a MIPI CSI-2 hardware interface for a digital camera (used for either stills or video). The Mobile Industry Processor Interface (MIPI) standards group that defined this Camera

Serial Interface (CSI) also defined the Display Serial Interface referred to above.

The type of camera that this connector is designed for (the BCM 2835 can deal with resolutions up to 20Mpixels) would be impossible to deal with in a timely manner on the ARM. The VideoCore, on the other hand, has (literally) won prizes for its ability to deliver 'the best camera phone in the world'.

Unfortunately the tuning and other software work needed to provide an interface to a Raspberry Pi camera from the VideoCore processors wasn't complete at the time of writing, but when one is released it will plug in to this CSI connector.

What to do next

Now you have Raspberry Pi up and working you can use it in many ways.

- You may wish to explore the small number of programs that are available from the desktop.
- You could open a command window and, with help from the section about Linux, explore what the commands do.
- To keep the size of the SD card down a very small set of software is installed initially. There's a huge range of other programs available to download from the Internet. We explain how software is installed in the section about Linux.
- You can learn a programming language. Later sections have some help getting started with Python, but C, Bash and ARM assembler are also beckoning.
- You could find out how to use Python to write games. There's a brief introduction to a simple game later on.
- You can use Raspberry Pi to control your own hardware. We've seen what you can plug into it and later sections give you more detail about individual interfaces.
- You can develop a project incorporating your own software and your own hardware. Later sections go through a menu of 'software recipes' which describe what parts of Linux can help with things like writing your own Web pages to control your software, and making your programs run regularly on their own. Other sections provide 'hardware recipes' that help you add standard devices, like WiFi adapters, USB memory sticks and so on.

The CSI connector.

03.
How to cook

HOW TO COOK
Introduction to programming languages

Processors spend their time following lists of instructions which are represented as data in memory.

Machine code

In the earliest computers programmers supplied this data by replicating the bits in the data exactly, perhaps as holes in a paper tape or as a series of switches on a console. Entering 'machine code' in this way was not only enormously tedious, it was also highly error-prone. To operators who hadn't written programs themselves the code was essentially meaningless, but just one switch or punched hole out of place could cause unexpected behaviour when that instruction was executed.

 Bugs

Errors in programs – which are always unexpected to programmers – are called 'bugs'. This term had been used in America to describe technological glitches since well before the invention of computers, but was originally a description of the irritation they caused, in the sense of the inconvenience 'bugging' you.

Rear Admiral Grace Murray Hopper, who amongst other things was responsible for the early and disconcertingly verbose COBOL programming language, may have helped develop the modern meaning of 'bug'. In 1947 an error in her program was traced to a real moth that had become stuck in the relay of the computer she was using at Harvard. Following that she would talk of 'debugging' problems found in programs.

Assembly code

Many of these problems were solved with the idea of an assembler program. This program translates a more person-friendly representation of the instructions ('assembler code') into machine code. Because this involved no human intervention at least one cause of error was eliminated.

The first machine code representations for people to use were very simple: they consisted of a series of lines of text laid out like a table, with each line describing one machine code instruction and perhaps giving a human readable name that could be used to refer to the position of that instruction. These location 'labels' were then used elsewhere in the program when needed. Programs represented this way could be recorded and reused later wherever they were due to be located in the computer.

The actual set of instructions an assembler understands will depend on the kind of processor it makes machine code for. Someone who writes an assembly code program that can play the game of noughts-and-crosses on, say, an Intel x86 processor can't use her program on an ARM processor because the instruction set is different. In the days when almost every computer had a different instruction set, submitting her program to the American Journal of Tic-Tac-Toe in assembler code would be almost pointless, because few others would understand it.

Programming languages

Proper high-level programming languages take one further step towards describing the method (or 'algorithm') that's to be used rather than the precise set of instructions, and, in principle, allow the same program specification to be used no matter which instruction set is available. Although they're still expressed using lines of text, as they are in assembly code, very little else is similar.

 High-level and Low-level Languages

The more difficult it is to determine which machine code instruction might be used when writing in a programming language, the 'higher level' it's said to be. Machine code and assembler code are very low-level languages, but even among other programming languages some are more abstract than others. The highest level languages concentrate on describing the problem and working out the solution themselves, rather than having the programmer specify it directly.

Animating programs

Once a program has been written there is, in fact, more than one way to bring it to life and 'run' it.

Compilation

We've really already come across one way to run a program when we described an assembler: a translator reads the program and decides which machine code would best represent it.

If the translation was correct the machine code program created by the translation will do exactly what the programmer

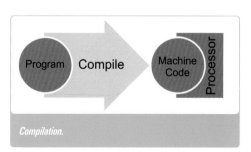

Compilation.

Programming Language Names

In the early '60s the Mathematical (now Computing) Laboratory at Cambridge University invented a programming language it initially called CPL after 'Cambridge Programming Language', and then, after the project was joined by London University, 'Combined Programming Language'. Never used very widely, a simplified 'Basic' version was derived by Martin Richards, called BCPL.

Kenneth Thompson stripped this language of all he could do without to create one simply called 'B' that was used when writing the first versions of the Unix operating system. Limitations then drove Thompson and Dennis Richie to develop a typed successor that they used to rewrite the central part of Unix (its kernel). Being a successor to 'B' they called it 'C' – now one of the most widely used of all programming languages.

intended when its instructions are executed by the processor. Again this translation is normally accomplished by a program (quite a complicated one in this case). For programming languages this program is called a 'compiler', and the translation process is called 'compilation'.

Interpretation

Given that compilers understand a programming language and produce machine code to implement it, it must be possible for a similar program to perform the task that the machine code would have undertaken immediately. Such programs essentially behave like a processor themselves: accepting and executing the instructions written in the programming language, instead of a processor doing the same to machine code instructions. The program that decides

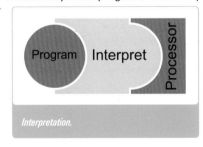

Interpretation.

the meaning of the program and then itself performs the actions it specifies is called an 'interpreter'.

Running a program using an interpreter will usually be slower than compiling it and then running the machine code. This is because the interpreter is doing more work than the processor: in addition to running the program it also has to decide what it means as it goes along. Being ten or even a hundred times slower isn't uncommon, but speed isn't everything: interpreted languages are usually more friendly to the programmer, and so easier to write, and they're easier and faster to debug.

Raspberry Pi languages

Your Raspberry Pi will, in fact, allow you to experiment with quite a lot of languages – they can be installed over the Internet quite simply. They're very often backed by an enthusiastic group of geeks who provide documentation and help through the Web.

There are only a few languages that are particularly important

for this book, though: Python, C and Bash (as well as ARM assembly code if you count it as a programming language). These all have many common features that they share with a pretty large group of useful languages. In the rest of this section we'll outline some of these features in a way that will help when we come to introduce some of them more explicitly later.

Two of the most important aspects of a language are: how it allows you to structure your program; and how it allows you to structure your data. In both of these there are three important ways to combine things – concatenation, alternation and repetition:

- *Concatenation* is an arrangement of two or more things following each other to make something bigger.
- *Alternation* is an arrangement of two or more choices.
- *Repetition* is an arrangement that allows the same thing to be used many times.

Names

It's a matter of convenience to a programmer that things in a program can be named so that they can be referred to later. Sometimes quite unusual things can have a name, but the most useful things to name are program structures and data values.

Program structures

If your Raspberry Pi were connected to a knitting machine an example of a program structure that could be named might be a list of instructions that describes how to knit a row from a knitting pattern. Naming these instructions (let's say 'fluffyrow') would make it easier to describe how a particular panel design is knitted, because you could simply say something like 'knit a "fluffyrow" for the next four rows' instead of spelling out the same instructions four times.

As well as making repeated references easy, naming things has another significant advantage: it can make the overall structure more understandable. If following the 'fluffyrow' instructions results in something that actually does look fluffy then naming it and using that name gives a reader some information she might not have had otherwise. More profoundly, consider the difference between instructions that say 'fluffyrow, fluffyrow, fluffyrow, fluffyrow' and the equivalent instructions that simply repeated the 'fluffyrow' stitch-by-stitch instructions four times. The reader may easily not spot that the same instructions are being repeated, or – amongst so many stitches – even be able to see that they describe four rows. In a very real sense the use of the name helps a reader see the structure underlying the instructions.

A good programmer will use names meaningfully and often – and not only to avoid duplication.

Data values

In an arcade game where you move a gorilla to different locations on a screen, a picture of the gorilla holding the hammer above his head; another holding it horizontally; and a final one holding it at his feet, might be three examples of data values that might be named (let's say 'up', 'heave' and 'splat'). Using these names makes it easy to describe how to animate a falling hammer (show 'up', then 'heave', then 'splat') and a rising one (show 'splat', then 'heave', then 'up') without having to describe the same picture twice.

As with program structures, naming data values can be used to bring out meaning. A knitting program might use the number 84 many times. If this data were named 'stitches_per_row' its intent might be much more obvious, even though the programmer will have to type more while writing a program. (By the way, most programming languages will allow you to put the underline character (_) in a name but won't let you use a blank space.)

Variables

A variable can be thought of as a pigeon hole that you can put values into which you can look at later. Allowing you to name them is one of the most important things a programming language does for you.

Programming languages can be pretty inventive when it comes to things that can be put in variables. These include simple types of data such as various kinds of number, strings of characters and Boolean: most languages can store these in variables.

More advanced programming languages will allow you to make up your own type of data to put in a variable, and many programming languages allow you to place a program structure in a variable. (For example, a knitting program might use a variable to remember an arbitrary list of instructions that it'll obey after completing every row.)

Programming Language Names

Lord Byron could never have known the contribution his daughter was to make to computing science. As a mathematician, Augusta Ada King, Countess of Lovelace, was interested in Charles Babbage's pioneering mechanical 'analytical engine'. Being too complex to be built (although a previous design has been recreated in the London Science Museum) this machine – the first true computer – was the subject of only research papers during his lifetime. It was in Ada's notes on one of these papers (which she was translating into English) that she described an 'algorithm' that would make the machine perform a particular function. This is taken by most people to be the first true program, and Ada, therefore, the first true programmer.

The comprehensive 'Ada' programming language was named in her honour and for a long time was required to be used in all programs in the British and American military.

Types

In some languages (like Python) any values can be placed in any variable, even if you'd think it doesn't make much sense. For example, if you became confused between your gorilla game and your knitting program you might put a series of three pictures for an animation in a variable called 'stitches'. Assuming that your knitting program counts the number of stitches by adding one to the value in 'stitches' every now and then, you can expect the language to get a bit upset when it's asked to add one to a group of three pictures. A stitch count

and an animation are really quite different types of thing: what you can do with one you can't necessarily do with the other.

To help prevent this kind of nonsense other programming languages (like C) oblige you to declare what type of thing you can put in variables before you use them. Compilers for these languages are usually clever enough to work out what sort of thing you're trying to put in a variable and will complain when you try to do something like add a stitch to an animation. On finding an error like this they'll often refuse to generate a program – which gets around the problem of working out what it would do if it ran.

Languages which don't provide the programmer with a way to describe the types of value that variables must contain still keep track of the type of every value, but they can't warn you that you're doing something nonsensical until they actually attempt to do it. Keeping track of the types of each value at 'run time' is an additional expense and makes programs written in those languages run a little slower.

Languages that support them need to have a way to describe the 'type' its values must have. These descriptions can get quite complex and may require a little language-inside-a-language of their own. Just as programming languages allow you to name program structures and data values, these languages will allow you to name the type description.

Program structures

Every programming language has its own syntax that it uses to describe various program structures. Here we're going to look at what these structures strive to achieve, without going into any particular syntax. We'll look at the syntax that Python uses, as a real language example.

Concatenation

In most languages the idea of joining program structures together is so natural that programmers barely notice it. It's normally simply achieved by listing one instruction, or other program structure, after another. The resulting sequence is executed one by one until it's complete.

Like all the other program structures it's normal to be able to use a concatenation of instructions wherever any instruction might be expected.

Alternation

In alternating program structures a choice is made among instructions and only one is executed.

The most common alternating program structure ('if') uses a Boolean value ('true' or 'false') to select which of two alternatives is executed. For example, a value from a Boolean variable 'bored' might be used, so that if it's true instructions to watch television are executed, and that if it's false instructions to knit another row are executed.

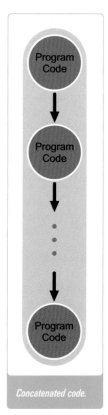

Concatenated code.

In a less common structure ('case') each of the alternatives are labelled with a value that must be matched if the alternative is to be executed. So, for example, if we have a joy-stick variable that can take on either of the values 'up', 'down', 'left' or 'right' we might use a 'case' structure where the instructions labelled 'up' move a gorilla up a game screen, those labelled 'down' move it down, 'left' left and 'right' right.

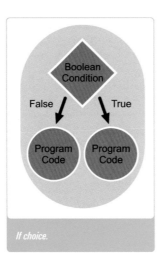

If choice.

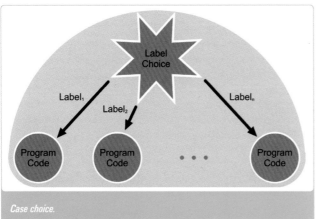

Case choice.

Programming Language Names

Mathematicians will recognize the terms 'scalar', 'vector' and 'matrix' which describe a single number, a row of numbers in a column and a two-dimensional grid of numbers; and they'll know that these can be combined in a few ways, just as ordinary numbers can, using operations like addition and multiplication.

The layout of numbers doesn't have to stop in the second dimension with a grid, and Kenneth E. Iverson invented a notation that could be applied to arrays of any dimension. The number of operations it supported, each represented by its own symbol, was so large that special-purpose keyboards had to be used to type the notation. The book he wrote to describe the notation was called *A Programming Language*, and that name stuck: the language is called APL.

Repetition

In repeating program structure an instruction is executed over and over again according to some criterion.

In a 'while' program structure a Boolean condition is evaluated to determine whether the instruction should be executed. For as long as the condition is 'true' the instruction will continue to be executed, stopping only when it becomes 'false'. For example, we might use our 'bored' variable to supply the condition and repeatedly execute instructions which take a sip of tea and then update 'bored' to either 'true' or 'false'. The result would sip tea until bored was 'false'.

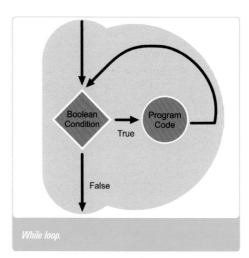

While loop.

A very similar program structure 'dountil' does almost the same but executes the instruction before evaluating the condition for the first time. Effectively it's the same as the 'while' construct except that the instruction is always run at least once (which isn't the case for 'while').

The last common repeating program structure is 'for', which needs the specification of a variable and a sequence of values for the variable to take. The repeated instruction is executed once for each of the values once it's been placed in the variable. For example, in a game a sequence of neighbouring squares inside a blast radius might be provided as a 'for' sequence and given to a variable 'splatted' along with a concatenation of instructions that destroys any player found on the square referred to by 'splatted'. In effect this would destroy any player in the blast radius.

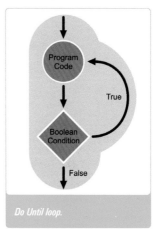

Do Until loop.

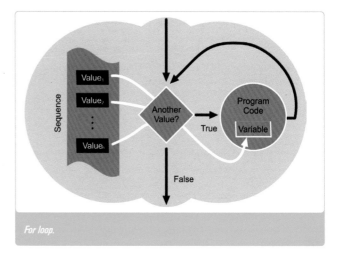

For loop.

Data structures

As with program structures, we'll wait until later in the book before describing the way that our languages express these different kinds of structure.

Programming languages usually have:

■ Some way to describe a data *value* with the structure you want; and (if it supports types)
■ Another way to describe the *type* of a value that a variable can contain.

The type of a value can be specified using the concatenation, alternation or repetition of other types (in a clear parallel to the way a program structure can be specified). Type specifications, however, aren't necessary in the languages we'll describe in this book.

A value can have a structure that has separate parts to it and these kinds of structure can be distinguished by the way those parts can be retrieved.

Structures whose parts have names

An obvious way to create a structured data value from other values is to give each part its own name. For example, if you want a value to represent a person you might choose to construct it out of three other values: a number to be called 'age', a Boolean to be called 'gender', and a string of characters to be called 'name'.

If you have a knitting program that can knit jumpers for clients you might represent a client by joining together a value like the one above, calling it 'person', and adding three measurement numbers, called 'chest', 'waist' and 'arms'.

In Python this kind of value is called a dictionary, whereas in C it's called a record or a 'struct'.

Structures whose parts are numbered

Another way to build up a data value is to number several other values. For example, if you have data values you use to represent your MP3 music tracks in an album you could represent an album in a combined data value that successively numbers each MP3 with the number of its track position.

In Python two types of structured value like this are available, one called a tuple and the other a list. Bash and C have similar structures called an array.

Comments

Perhaps one of the most important constructs that's available in every programming language is one that allows some program text to be completely ignored.

It's an unfortunate fact that the purpose of a program is often not readily apparent to a reader who wasn't its author. Sadly, after a few months it may not be very obvious to the author either. For every programmer who has a sharp enough insight into another coder's mind to uncover its point there's another who has a sufficiently perverse and convoluted mind to write programs that'll make an exception.

Called a 'comment', this ignore-all-of-this-text construct provides a way for the author to at least provide hints to a future reader (whether herself or others) as to what on earth the code was intended to do.

HOW TO COOK

Introduction To Operating Systems

Computers are virtually everywhere. Your laptop is a pretty obvious example, as are the computers in your phone and your car's GPS system, and lately the ones in your 'set-top box', DVD player and TV too. In fact they're more pervasive than that: there are little ones embedded in all kinds of places – your watch, your new digital radio, your hi-fi, your washing machine and even the clock on the wall. Your kitchen and car are probably packed with them.

The main difference between the really obvious computers and those embedded in all kinds of other devices is whether you get to choose what program they run: embedded computers typically have a fixed program that never changes; your laptop (you might think) runs the programs you ask it to. In either case it's the ability to run programs that make computers useful, and operating systems are there to make that possible.

Threads

Most programs describe how to proceed with one thing taking place after another has finished – as if each step were connected by a thread. Nonetheless, while you have your word processor open ready to type a letter you perhaps have your browser open on Facebook while your favourite MP3 plays in the background. If each program is following its own thread, the computer must be supporting at least three threads at once.

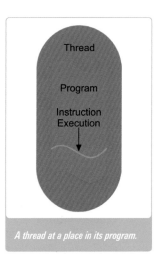

A thread at a place in its program.

Providing threads is perhaps the most significant thing an operating system does. Like the three Fates of ancient Greece, an operating system spins the thread of life, allots and measures it, and finally cuts it. The operating system: gives threads all they need to live (a program, some memory, a user and a processor to run on); keeps them safe from the other threads; and, punishes them for their indiscretions.

From time to time the program a thread runs might create child threads. Operating systems often group together families of threads created this way into a construct called a 'process' (or sometimes a 'task') so that they can provide things to the whole family rather than to each individual. Operating systems also provide all kinds of services to threads by giving them access to other threads and hardware devices. We'll bundle together all of these things under the general term 'resources'.

The kernel

When an operating system has given a resource to one process, what's to stop another process from using it instead? One of several ways that operating systems solve this problem (if they bother at all) is to monitor all resource use in a special program called a 'kernel' which can't be tampered with. All of the instructions that are part of the kernel run in a special mode provided by the processor that allows them privileged access to the computer's resources. The instructions in user's threads don't have this privilege and therefore have to ask the kernel to do some things on their behalf.

Essentially this splits the programs in the computer into two: parts of the privileged kernel (which is said to run 'in kernel mode') and programs in the unprivileged processes (which are said to run 'in user mode').

Running the kernel

The kernel is just a big program, and in order for it to do its job it obviously needs to run sometimes. There are two ways that it can be made to run: 'system calls' from processes and 'interrupts' from hardware.

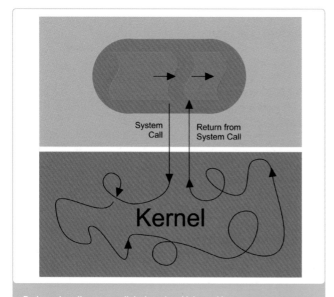

During a thread's system call the kernel could do anything.

System calls

User-mode programs have to use a special processor instruction to utilise features of the kernel, sometimes called a system call. This instruction stops the thread and starts running the kernel in its more-privileged mode.

When the operating system has finished doing what was requested of it it uses another instruction that 'returns', causing the user-mode program to start executing again at the next instruction in its thread (just after its system call), but in the more restricted user-mode.

Interrupts

The processors on a chip are connected to any number of hardware devices. Any of these can raise an electrical signal to ask the processor for attention. These signals are a bit like a doorbell: there's an expectation that you'll 'interrupt' what you're doing and go and investigate.

A processor responds to an interrupt by stopping the thread it was executing and running a program (an interrupt handler) in the kernel, to see what's wanted.

In a short while the kernel will have done all it needs to service the interrupt and it'll then return to the thread that was interrupted. It does this by using another special processor instruction that, like the system call return, first restores the thread's unprivileged state.

Devices

There are many interrupt handlers in the kernel, each looking after a particular kind of device. Typical devices are mice, keyboards, disk drives, memory sticks, network interfaces, graphics cards and any of many, many components that you can find on the same circuit board as the processor or even in the same chip as the processor.

The interrupt handler and the other kernel programs used for the system calls that allow processes to access a particular type of hardware are called its 'device driver'. If you invent something of your own, such as an electronic mousetrap or an alarm clock based on over-inflating balloons, it'll probably need its own driver.

Like applications, device drivers can be supplied either as part of an operating system distribution or added later.

Scheduling

So, the kernel is entered both when a system call is made by a thread, and when a device causes an interrupt. In both cases a thread will be suspended while the kernel does what it has to.

Because there can be many more threads than processors there are usually many suspended waiting to be run again. There's no reason that the kernel has to resume the same thread that was last suspended. In fact it'll usually choose, or 'schedule', another thread. All the threads need to run at some time, after all.

For example, the kernel might program a 'timer' device to deliver an interrupt 100 times a second. This will mean that the kernel is entered regularly and can decide, at least that often, which threads ought to be running and make changes as necessary. With different scheduling rules it might choose the thread that was run the longest ago, or it might choose the thread that was the most important, or it might just choose the 'next' thread.

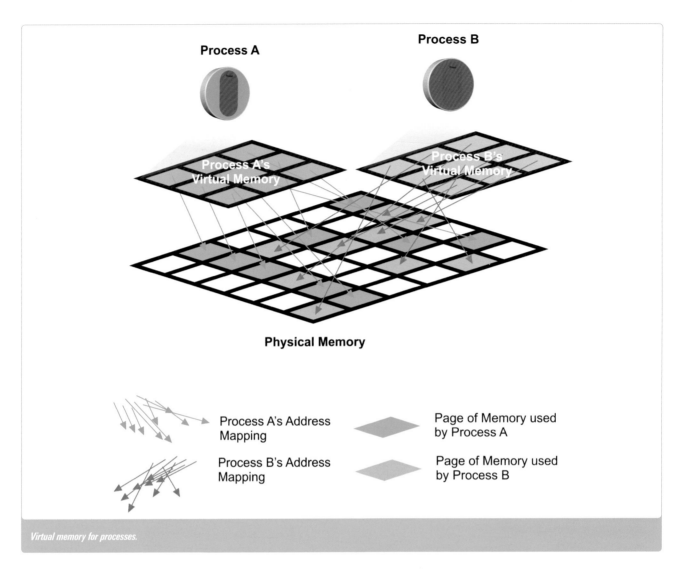

Process A

Process B

Process A's Virtual Memory

Process B's Virtual Memory

Physical Memory

Process A's Address Mapping

Process B's Address Mapping

Page of Memory used by Process A

Page of Memory used by Process B

Virtual memory for processes.

Virtual memory

Special hardware called a Memory Management Unit (MMU) is often available that can determine the parts of the computer's memory that are available to a process.

The memory in the computer that the MMU can use is called 'physical memory', but since not all of this can be used we need a different term for the memory that's actually seen by a process: 'virtual memory'.

When a running thread uses some (virtual) memory, say at location zero, the MMU might actually provide memory that has a completely different address in physical memory.

Because the kernel can reprogram the MMU before resuming a thread it can give every thread a completely separate set of physical memory areas (a second thread might use a different area of physical memory for its virtual memory at location zero, for example).

The kernel might choose not to put any memory at all in some areas of virtual memory. The collection of virtual memory areas that a thread can sensibly use is called its virtual 'address-space'. Usually the kernel will arrange that all the threads in the same process share exactly the same address-space.

Swapping

Some kernels (including both Linux and Windows) do something very clever with MMUs that allows there to be more virtual memory than there is real physical memory!

To do this the kernel needs somewhere it can store data, usually an area of disk, called 'swap space'. When all of the physical memory's been used up (and sometimes even before that happens) the kernel looks at its collection of threads to see which are the most inactive. (Perhaps that word processor, open for writing thank-you letters, isn't very busy.) It then chooses some of its physical memory, writes ('swaps out') it out to its own part of the swap space, and makes that physical memory available to use.

Like other pieces of hardware the MMU can generate interrupts, and it'll do this when the processor tries to use an area of virtual memory which has no physical memory behind it. Our poor word processor will be completely unaware that some of its memory has gone missing and, once its user starts typing 'Thank you for the lovely socks', it may well try to use some of its virtual memory that isn't there, either by executing program code, reading from memory or writing to memory. When this happens the MMU will cause an interrupt, and, as we've already said, the kernel will suspend the (word processor) thread and run some program to handle the interrupt. The handler program will find some physical memory and read ('swap in') into it the copy of its virtual memory it wrote there (when it was swapped out). Then it'll update the MMU so as to use the new physical memory for the missing virtual memory and resume the thread that caused the interrupt.

The amazing thing, as far as the word processor is concerned, is that it won't be aware of what's happened at all. It simply executed its program instruction by instruction continuously, unaware that between one instruction and the next there had been a long delay while the kernel did the swapping it needed.

So the memory that the word processor 'thinks' it has is, in a sense, not 'real' … it doesn't exist in its memory map where it thinks it is, and it might not even exist in physical memory. Perhaps you can see why the addresses it uses are called 'virtual'.

File systems

In addition to lots of application programs, the kernel and device drivers an operating system distribution will usually include many kernel- and user-mode libraries that fill the gap between the simple things that a device driver will do and the complicated things users normally want to use them for.

For example, the device model in both Windows and Linux is based on the old Small Computer System Interface (SCSI) disks that Apple made popular. It splits the disk up into fixed-size blocks of bytes (usually called 'sectors'), numbers them and allows you to read or write whole sectors starting at a specific number.

Placing your collection of photos in this (very) long list of blocks and finding them again would be a nightmare. This is the impracticability that file systems address (either in the kernel or user-mode libraries).

Files

Files are just general-purpose containers for long series of bytes. They're the first thing that a file system implements using the long list of sectors a storage device might provide. The most basic property of a file is its size – a file can be any number of bytes in length (and doesn't have to be a whole number of sectors long). The file system remembers a file's length and which sectors on the device are used. Because mechanical disks read and write sectors that are next to each other faster than those that aren't, a file system will try to arrange files so as to use big stretches of consecutive sectors.

Directories

The next thing that a file system must provide is some way to name files. In Linux the names are kept in 'directories', but Windows sometimes calls them 'folders'.

A directory has one entry per file, usually with its name and something that indicates where the file is. It might also include other attributes of the file, such as the permissions different users have to read and write it.

In some file systems a directory is really just a file put to this special use, in others a directory is a different kind of structure altogether. In both cases, though, directories can usually be given a name just as a normal file can.

Each disk contains a unique 'root' directory through which all other files can be found. It isn't necessary for all the files on a disk to be named in the root directory directly, because the root might contain references to other directories which might contain more files, or directories, and so on. You can think of the root directory as the stem of a plant and each directory as a place where it branches in separate directions, one branch for each of its entries. The whole collection of files will look a bit like a tree with every final 'leaf' being a file. This is the widely adopted simile that gives the 'root' directory its name.

File path names

The final thing that a file system has to support is a file path name. A file path allows a specific file to be identified, starting at a disk device's root directory. It includes the names of successive directories locating the one finally naming the file.

For example, suppose that you have nothing on your disk except your maternal family tree represented by directories named after women holding files for each of their male children and directories for each of their daughters. My grandmother Laura had a daughter, Lynn, who had a son, Gray. If Laura is in the root file system a name that first names the disk and then gives the names Laura, Lynn, Gray would be sufficient to identify the file representing me.

Different file systems can choose different ways to divide up the parts of a file name. Windows uses disk names with a single character in them, like 'C', and separates this part from the rest with a colon (:) and backslash (\), and then separates different parts with the backslash character (\). If my family tree were on disk 'C' my file would therefore be called 'C:\Laura\Lynn\Gray'.

Linux always puts the file system on a disk inside other file systems, just as if they were another directory. It uses slash (/) as the name of the root and then separates each subsequent part with another slash (/). If the name of the disk was '/family' my file would be called '/family/Laura/Lynn/Gray'.

Other file systems (such as RISCOS) have used a colon to separate the name of the disk from the file, and then a full stop to separate remaining parts. That would have resulted in my file being called 'family:Laura.Lynn.Gray'.

Partitions

The blocks of data used by a file system normally come from a long sequence of blocks with consecutive addresses. On a disk this sequence doesn't have to use up the whole disk: there may be several stretches of the disk each containing sequences of blocks holding different file systems. These divisions of the disk are called 'partitions'.

Most partitions will be laid out using a file system, but not all of them will. For example, there's a special type of partition that itself contains further partitions. Other types of partition provide a set of sectors used for swapping to, and still others simply identify empty areas of the disk.

It's normal for a disk to begin with a tiny amount of data called a 'partition table' that describes how many partitions there are, where they occur on the disk, and what they're used for. In many cases it'll describe only one partition that uses up the whole disk, and that partition will be used for the main file system.

On Raspberry Pi the SD memory card is normally used as if it were a disk and has two or three partitions.

HOW TO COOK
Introduction To Python

*"First shalt thou take out the Holy Pin,
    then shalt thou count to three, ..."*
(Python program to operate the Holy Hand Grenade of Antioch,
Monty Python and the Holy Grail)

During the 1980s the holy grail of operating system design was to create one that could manage any number of connected computers as if they were just one. When the operating systems guru Andrew Tanenbaum created Amoeba, a working distributed operating system, at the Vrije Universiteit in the Netherlands, he probably wouldn't have thought that the language designed to administer the system is now used far more than the system itself. Python was born because Guido van Rossum didn't like the Bourne shell traditionally used for such purposes and wanted a clear, easy-to-read language that he could extend without difficulty.

Great things about Python

Python is typically interpreted on a virtual machine. This means there's normally no explicit 'compilation' step needed before a program can be run. In fact you can type a line of program into a Python interpreter and see the result execute immediately.

```
pi@raspberrypi:~$ python
Python 2.6.6 (r266:84292, Dec 27 2010, 21:57:32)
[GCC 4.4.5 20100902 (prerelease)] on linux2
Type "help", "copyright", "credits" or
"license" for more information.
>>> print("Hello, world!")
Hello, world!
>>>
```

Python has few unnecessary language features and a simple layout. The structure of your program is partly dictated by how many spaces you use in front of each line. Nonetheless, it supports many of the features needed to support the best programming practice.

It's now been adopted so widely that there are a huge number of standard libraries available which do all kinds of useful things requiring virtually no work by the programmer.

Best of all, most implementations are 'open' and this has helped make it very popular. If you invest time learning how to use the language, you'll find that you can write your programs almost everywhere: Python is freely available on all versions of Linux, and Windows, and Mac OS, and many other operating systems. On systems that don't have Python you can install it yourself with no worries about fees or licensing. Furthermore, Python is not a cut-down language intended only for teaching: you can use it even for large and complex projects. On top of that it'll continue to be supported by a community of enthusiasts for a long time, not simply for the lifetime of a company that sells it.

And, most importantly, it was named after *Monty Python's Flying Circus*.

Overview

In this section we're first going to see how each of the things we described in the section about programming languages applies to Python; then we'll go through a short program as an example. It won't make you an expert in Python, or even show you all of the language, but it should allow you to start experimenting yourself.

Preparation

It's more fun to try out examples as they're described. In order to do this we're going to use two different environments: the command line for simple examples, and the 'Geany' development environment for more complicated ones.

For this Raspberry Pi should be started with a graphical desktop with a terminal (command-line) interface as was described in the section about the first-time Raspberry Pi.

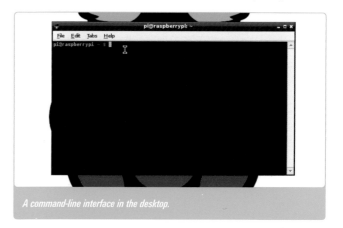

A command-line interface in the desktop.

Interactive Python window

To try out simple Python programs and commands just type the following into a command-line window:

```
python
```

(as we did above). Following that, anything you type in is treated as a Python program which will be run immediately.

You can return to the command-line window by typing Ctrl-D (for which you must hold down the 'Ctrl' and the 'D' keys at the same time).

You can use the command-line interface to try out the examples in the first part of this section by first typing the command 'python'.

```
pi@raspberrypi ~ $ python
Python 2.7.3rc2 (default, May  6 2012,
20:02:25)
[GCC 4.6.3] on linux2
Type "help", "copyright", "credits" or
"license" for more information.
>>>
```

Following that everything you type will be prompted for by the characters '>>>' and treated as a python command. If you just type in an expression Python will evaluate it and print out the result:

```
>>> 6*7
42
```

In Python, as in most programming languages, addition, subtraction, multiplication and division are all available and are evaluated just as you were taught in your maths class. The only difference is that although the addition and subtraction symbols are the familiar '+' and '-' characters, the multiplication and division symbols are '*' and '/', not 'x' and '÷'.

While on the subject of sums, don't forget the lesson when your teacher told you that multiplications and divisions are always done before additions and subtractions.

```
>>> 6+5*7+1
42
>>> (6+5)*(7+1)
88
```

Geany

Geany is a program that organizes one or more files containing programs into separate 'projects'. The program allows these programs to be edited easily and provides a simple way to try them out. It can be used both for Python and for C programs (which are addressed in a later section).

Once you have a graphical user interface Geany can be run by clicking on the icon in the bottom left of the screen and selecting 'Programming' followed by 'Geany'. If you don't see this on your Raspberry Pi you may need to install it first from the Internet. To do this type this into a command-line:

```
sudo apt-get install geany
```

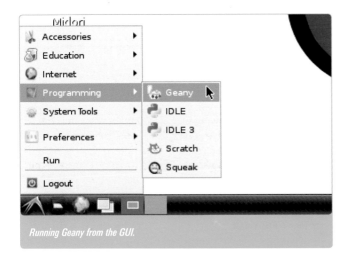

Running Geany from the GUI.

(You'll have to have a network connection first.) Once this completes you should find that Geany is available as above.

You might be interested to hear that Python and Geany are both available for your Windows computer (if you have one), so you can try out programs there, as well as on Raspberry Pi, if you want.

You'll find Geany at

```
http://www.geany.org/
```

where there are instructions on how to download and install it. Similarly you can find Python at

```
http://python.org/download/
```

Look for a Windows download.

Using Geany

Geany assumes that your program may one day become too large for a single file and so provides a directory where all the files in your program can be held together. Initially, though, you'll create only one program file.

Because you might be working on more than one program at once Geany also remembers which 'project' you're working on and allows you to switch between one and another. This means that the first time you use Geany you need to:

■ Create a new project to work on.
■ Create a new Python file to hold your program.

The following provides a step-by-step illustration of how this can be achieved together with writing a trivial program and running it. Running Geany for the first time gives:

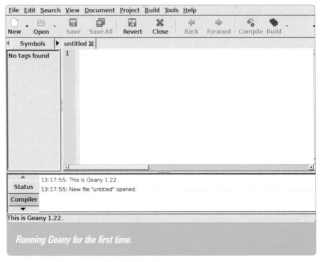

Running Geany for the first time.

First create a new project using the 'Project' menu item.

Creating a new project from the menu.

This will ask for the name of the project. Type in a name and click on the 'Create' button.

You may get a warning that the project directory Geany needs to create does not exist yet. Just click on the 'OK' button to allow Geany to create it for you.

To create the first file in this project use the small symbol to the right of the 'New' toolbar button and select 'main.py' (the other options create new files in other programming languages).

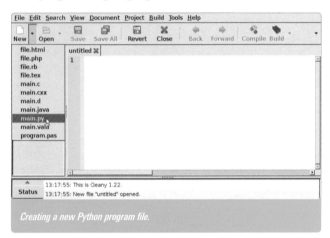

Creating a new Python program file.

The new file Geany creates for you will already have a number of lines in it, including a large comment that will result in your program being freely available to others under the GNU Public License (GPL) if you don't remove it. (You are quite free to delete it if you wish.)

You can use the scroll bar, the down arrow key, the 'end' key on your keyboard or your mouse wheel to move to the bottom of the file. There is an area where you can write new lines of program. To test Geany out type this line after 'def main():'

```
print("Hello world!")
```

then click the 'Save' button on the toolbar.

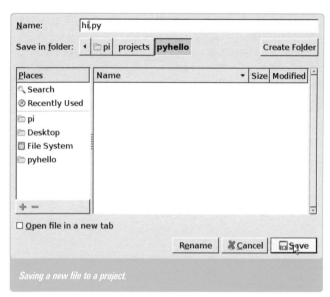

Saving a new file to a project.

To run the new program simply click on the 'execute' button on the toolbar. This will result in a new window being opened in which the output of the program can be seen.

Testing the new program.

Now that the program has been written, saved and tested the file can be closed using the 'Close' button on the toolbar.

Closing the Python program file.

At this point Geany could be closed. If it were closed the same project would automatically be loaded on the next run. To prevent this close the whole project using the 'Project' menu:

Closing the whole project.

Because the file has not been saved before you are given a chance to name it yourself. Typing a new name will simply replace the high-lit 'untitled' text. For example if you wanted to name the file 'hi' type the name and click on the 'Save' button:

Comments

A Python program is built up from separate lines, each of which can end in a comment. A comment starts as soon as a hash character (#) is typed and, although they can, lines don't have to contain anything except a comment. So one of the simplest programs in Python is:

```
# This program does
# nothing at all
```

The 'Hello world' example above could be commented using:

```
Print ("Hello world!") # print a greeting
```

Names

In some ways, typical of interpreted languages, Python naming is simpler than other programming languages. Python names can be given to program structures and variables but not types or values. (Types can't be named simply because the user never has to specify one.)

Names can contain alphabetic characters and digits as well as the underbar character (_), although (as is usual in many programming languages) they can't begin with a digit. You can use upper- and lower-case characters in a name. The following are all valid and different names: c3, Count, count, Holy_Hand_Grenade_Count.

Program structures

A Python function definition is a convenient way to name a program structure for later use. For example, the very simple program that just writes 'Hello world!' to the screen might be named 'Greeting', like this:

```
def greeting(): print("Hello world!")
```

To understand better how this works type it into an interactive Python window and then, on the next line type

```
greeting()
```

You should find that 'Hello world!' appears on the screen. The brackets after the name tell Python to execute the instructions that 'greeting' names.

Data values

Python doesn't really have a special way to name particular data values but programmers typically get the same effect by assigning the data value they want to a global variable which they never update. For example, before any functions have been written that need it a program might contain the line

```
Holy_Hand_Grenade_Initial_Count = 3
```

Wherever this important value is required in the program the name 'Holy_Hand_Grenade_Initial_Count' can be used instead.

Variables

Because it's a dynamically typed language it isn't necessary to indicate what type of value a variable will contain before it's first used. A new variable is created whenever an instruction

to put a value in it is executed. Naturally, if a variable has already been named the value will be placed in the variable that already exists. Thus in the two instructions

```
message = "Hello "
message = message + "World!"
```

the first will create a new variable called 'message' and put a string of characters (H, e, l, l, o and space) into it. The next instruction uses the value in 'message' and adds another string of characters to the end of it to form a value containing H, e, l, l, o, space, W, o, r, l, d, and finally '!'. This value is then placed into the variable already called 'message' (it doesn't create a new variable this time).

Program structures

Concatenation

In Python each line in the program begins with a number of spaces, which results in it being indented by a certain distance from the left-hand side of the page.

A block of instructions can be concatenated together simply by including them on successive lines with the same indentation (ie with the same number of spaces in front of each line).

The instructions in a block are executed one by one in the order that they're written. For example,

```
def prepare_lob():
    take_out_the_holy_pin()
    count_to_three()
```

joins together the two instructions that executes the instructions called 'take_out_the_holy_pin' and 'count_to_three' one after the other and names the result 'prepare_lob'.

Alternation

Python has an 'if' program structure but no 'case'. It tests a Boolean expression and executes one program structure if it's true and another otherwise.

```
if count == 3:
    print("The number of thy counting")
else:
    print("Thou shalt count to three, no more,
 no less")
```

The 'else:' and following instructions aren't required if there's nothing the programmer wishes to execute in this eventuality. It is, however, also possible to include the else with another 'if' test using 'elif':

```
def test_count():
    if count == 3:
        print("The number of thy counting")
    elif count == 4:
        print("Four shalt thou not count")
    elif count == 5:
        print("Five is right out")
    else:
        print("Thou shalt count to three,
 no more, no less")
```

Repetition

Referring back to the section about repetition in programming languages, Python has 'while' and 'for' program structures but not a 'dountil'.

While loops

A while loop takes a Boolean expression and executes its program structure repeatedly as long as the expression evaluates to true. For example,

```
while foe_not_snuffed_it():
    prepare_lob()
    lobbest_thou_thy_hand_grenade()
```

will execute the instructions called 'foe_not_snuffed_it' to evaluate a true or false value. As long as this is true, execute the concatenated program structure made first from the instructions called 'prepare_lob', then those called 'lobbest_thou_thy_hand_grenade', then all of this again, and keep repeating until 'foe_not_snuffed_it()' is false.

Because there's no limit on the number of times a 'while' can loop, this program structure can run forever. If it does end, however, we can be sure that 'foe_not_snuffed_it' has evaluated to false (and so, presumably, our foe is dead).

For loops

A 'for' loop takes a sequence of values to give to its variable and executes its program structure with the variable set to each of the values in turn.

```
def count_to_three():
    for number in ("one", "two", "three"):
        print(number)
```

In the above the value in brackets is a list of three text strings. The print statement is executed with the variable 'number' set first to '"one"', then to '"two"' and finally to '"three"'. Because the for loop is the whole of the program structure named, asking for 'count_to_three' to be executed will result in '"one"', '"two"' and '"three"' being printed.

Functions

A named program structure is called a 'function' in Python. We've already seen how they can be defined (using 'def') and how we can request that the instructions they name should be obeyed (ie how we can 'call' them) by placing brackets after the name.

Parameters

Functions also occur in most other programming languages and, as in those languages, they can be made much more useful by 'parameterising' them. Parameterisation involves defining some variables whose value will be set when the function is called.

For example, we've already seen how the 'greeting' function can print 'Hello World!', but suppose we want to have some instructions that provide a different salutation to the world, such as 'Good Day'? We could write a program that does something like

```
def greeting():
    print(salutation+" World")

salutation="Hello"
greeting()

salutation="Good Day"
greeting()
```

which uses a variable 'salutation' to hold the kind of welcome we want our 'greeting' instructions to make. This will print first 'Hello World', and then 'Good Day World'.

We can parameterise greeting() to provide an almost identical effect by adding the variable 'salutation' to the definition between the brackets:

```
def greeting(salutation):
    print(salutation+" World")
```

In effect this requires every call to 'greeting' to provide its own value for this variable. To ask for the instructions to be called it's now necessary to provide a value for 'salutation' between the brackets. This is the equivalent of the code above:

```
greeting("Hello")
greeting("Good Day")
```

Any number of parameters can be defined in this way. First of all they need to be named in the function definition (where they're called 'formal' parameters) – for example:

```
def greeting(salutation, audience):
    print(salutation+" "+audience)
```

Then later they have values provided for them in the function call (where these values are called 'actual' parameters):

```
greeting("Hello","World")
greeting("Goodbye","Cruel World")
```

Results

Almost all languages have a means to pass back a value which can be used where the function is called. In Python functions always return a value.

One way to see the value that a function returns is simply to print it out. Here we have a function that joins two strings together with a space between them and puts them in a variable called 'greet' but does nothing else. We can define it and then call it and print out the result:

```
def greeting(salutation, audience):
    greet = salutation+" "+audience
print(greeting("Hello","World"))
```

You'll see, if you do this, that the value 'None' is printed out. Every function in Python that doesn't explicitly say what its result is will return the special value 'None'. We use a special 'return' instruction to indicate what the result of a function should be. For example,

```
def greeting(salutation, audience):
    greet = salutation+" "+audience
    return greet
print(greeting("Hello","World"))
```

will now print out 'Hello World'.

No other instructions from the function will be executed once a 'return' instruction has been executed, so this instruction is often the last in a function.

Variable scope
You may be surprised by how this program behaves:

```
def greeting(salutation, audience):
    greet = salutation+" "+audience

greeting("Hello", "World")
print(greet)
```

If you try this in an interactive Python session you'll get a result like this:

```
>>> print(greet)
Traceback (most recent call last):
  File "<stdin>", line 1, in <module>
NameError: name 'greet' is not defined
>>>
```

The reason for this is that the variable 'greet' only exists while the instructions in 'greeting' are being obeyed. What's more, any variable that was already called 'greet' before the function is called is ignored once 'greet' has been given a value inside 'greeting'. For example, there are two separate variables called 'greet' in this program which aren't confused:

```
def greeting(salutation, audience):
    greet = salutation+" "+audience

greet="Boo!"
greeting("Hello", "World")
print(greet)
```

This will print out 'Boo!', not 'Hello World'.

Variables created inside functions like 'greet' are called 'local variables'. They exist only inside the function they're local to.

It's possible to specify that a variable inside a function shouldn't be local by using the 'global' instruction. The following will print out 'Hello World':

```
def greeting(salutation, audience):
    global greet
    greet = salutation+" "+audience

greet="Boo!"
greeting("Hello", "World")
print(greet)
```

It is the 'scope' of the variable 'greet' in function 'greeting' that differs between the way it's treated when it's local and when it's global. The scope of a variable determines the parts of a program where it would be used if its name were used.

Data structures
Because Python is dynamically typed every variable can accept a value with any data structure. Python determines the type of a data value from the way it's written.

Basic types
Python has a good number of basic types which are building blocks from which more complicated values can be built:

Type	Description	Examples
Integer	A whole number	0, -456, 17 0b00010001 (the binary value for 17) 0o21 (the octal value for 17) 0x11 (the hexadecimal value for 17)
Floating point	A number including a decimal fractional part	1., 3.1415, .5, 0.5 -1E6 (minus a million) 1e-6 (a millionth) 1.2e+3 (one thousand two hundred)
Complex	A floating point number including a multiple of the square root of minus one.	17+0j 0+2j, 2J, 2.0J 3.1415-0.5j

There are a small number of types that contain values constructed from a sequence of data with the same type. These are the most important:

Type	Description	Mutable	Example
String	Repeated characters	No	'spam' "spam" "sp" "am" (the following three lines:) """ spam """
Bytes	Repeated 8-bit values	No	b'spam' (the 8-bit values correspond to the ASCII encoding of 's', 'p', 'a' and 'm')
Bytearray	Repeated 8-bit values	Yes	bytearray('spam')

Unlike the other repeated types bytearrays can be written to. For example, this would print 'spam':

```
ba = bytearray('swam')
ba[1] = b'p'
print(ba)
```

Indexed types
There are a number of ways to combine values into a larger data value in Python. The combined value is either 'mutable' or 'immutable' depending on whether its parts can be changed.

When a Python variable contains a combined data value it's possible to refer to one of its parts using an index. An index can be a number to refer to the position of the part or a 'key' to refer to its name. Each indexed part can be treated as a value in its own right. For example, if 'films_year' is a collection of year numbers indexed by the name of the film you can use 'film_year["Holy Grail"]' as if it were a variable of its own.

```
film_year["Holy_Grail"] = 1975
print(film_year["Holy_Grail"])
```

The most important complex data types are:

Type	Mutable	Indexed-by	Example
List	Yes	Position	[1,2,3]
Tuple	No	Position	(1,2,3)
Dictionary	Yes	Key	{'first': 1, 'second': 2, 'third': 3}

The three values '1', '2', and '3' could be replaced by any other Python value (including a complex data type) and there need not be three of them in total – there could be any number, including none at all.

You might notice a problem with the representation of tuples of just one item: according to the example you'd expect a tuple consisting of only one value, say 2, to look something like '(2)', but this already has a meaning (it means '2'). The way to make it obvious that it's a tuple with '2' in it, and not just the value '2', is to include a final comma, but with no succeeding value, as in '(2,)'.

A position index is a positive whole number, with the first place being numbered zero (not one). Thus this would print the number '2':

```
tuple=(1,2,3)
print(tuple[1])
```

Although the examples of keys in a dictionary are all strings, a key can be any Python immutable data value (including tuples).

Mutable data values can have their elements changed after they've been assigned to a variable, while immutable ones cannot. So the following will print '[1,42,3]':

```
alist = [1, 2, 3]
alist[1] = 42
print(alist)
```

whereas

```
atuple = (1, 2, 3)
atuple[1] = 42
```

will just cause an error because you cannot assign a new value to part of an immutable data type.

Classes

In Python a 'class' can be thought of as another complex data type like a dictionary. The important thing about a class is that some of its parts are usually functions that do things to the class value. Such functions are often called the 'methods' of the class, and their first argument is always a class value.

Every class contains a special method called '__init__' which is called to set up its starting value. Here's an example class called 'Point':

```
class Point():

    def __init__(self, x, y):
        self.myx = x
        self.myy = y

    def distance(self, x, y):
        return ((x - self.myx)**2 + (y -
        self.myy)**2) ** 0.5
```

It has two methods '__init__' and 'distance'. Rather than using an index in square brackets to access its parts, as dictionaries do, the parts of a class are referred to using their name after a dot (.). When a new value of the Point class is created using the name of the class:

```
here = Point(1,2)
```

the resulting class value 'here' will have its two methods and gain an additional two parts 'myx' and 'myy', which '__init__' will have set to 1 and 2 respectively.

```
>>> here = Point(1,2)
>>> print(here.myx)
1
```

The function 'distance', which returns the distance between the point and another X, Y value, is referred to using a dot in the same way:

```
>>> here.distance(5,5)
5.0
```

Notice that we've provided two actual parameters, even though 'distance' has three formal ones. This is because Python always provides the class value ('here' in this example) as the first parameter of a method for you.

The great thing about classes is that they can effectively wrap up everything to do with a kind of object. With more methods our Point class definition could encapsulate everything we need in the representation of a Cartesian coordinate.

There's a style of programming that creates classes for a program's most important values which people call 'object orientated' programming. It's been extremely influential in language and system design over the last 50 years and has many features that make it desirable.

All of the basic types in Python have associated classes with a set of methods. For example, strings have an associated class and one of the methods in the string class is 'upper' – a function that upper-cases all the characters in the string. This is easy to demonstrate:

```
>>> "a string".upper()
'A STRING'
```

Documentation

Python has a way of providing documentation for your code as you write it. If the first line of a function definition isn't a line of code but a string, the text in that string will be associated with the function and can be printed out at any time. For example,

```
>>> def inc(n): "add one to n"; return n+1
...
>>> help(inc)
```

results in this text being printed out:

```
Help on function inc in module __main__:

inc(n)
    add one to n
```

Classes can be documented in the same way and when a class name is given to 'help', documentation for the class and all its methods is generated.

Example – Docker

To discover the quirks of a new programming language there's no real substitute for an actual example.

What follows is a very self-contained program, the simple aim of which is to insult its user in a manner similar to that of the rude Frenchman in the film *Monty Python and the Holy Grail* (something that no program can really accomplish adequately).

```
#!/usr/bin/env python
# -*- coding: utf-8 -*-
#
#   docker.py
#
#   Copyright 2012 Gray Girling
#
#   This program is free software; you can
#   redistribute it and/or modify it under the
#   terms of the GNU General Public License as
#   published by the Free Software Foundation;
#   either version 2 of the License, or (at
#   your option) any later version.
#
#   This program is distributed in the hope
#   that it will be useful, but WITHOUT ANY
#   WARRANTY; without even the implied warranty
#   of MERCHANTABILITY or FITNESS FOR A
#   PARTICULAR PURPOSE. See the #  GNU General
#   Public License for more details.
#
#   You should have received a copy of the
#   GNU General Public License along with
#   this program; if not, write to the
#   Free Software Foundation, Inc.,
#   51 Franklin Street, Fifth Floor,
#   Boston, MA 02110-1301, USA.
#
#
```

This program was written using the Geany integrated development environment. It provides some boiler-plate comments at the top of every program that, if you leave it in place, will mean it'll be covered under the terms of a GNU

General Public License (which is one of the types of 'free' licenses mentioned briefly in the introduction).

```
from random import *
```

The 'from' statement names a Python module and requests some of the things it defines. The asterisk character indicates that we're requesting everything that it defines.

The 'random' module is one of many predefined modules in Python. This one provides a number of functions to help exploit random numbers. In particular it provides the function 'randint',

```
number = randint(low, high)
```

which will return a random whole number whose value will be between 'low' and 'high' (including both of them).

It also provides 'shuffle',

```
shuffle(list)
```

which takes a list of items and shuffles them into a random order.

```
# Some lists of things we will use as parts of
   an insult:
relatives = ["mother", "father", "brother",
   "sister", "aunt", "nephew", "grandpa" ]
pets = ["dog", "elk", "pig", "horse",
   "parrot", "emu"]
bad_adjectives = ["wobbly", "gaseous",
   "unlovely", "flatulent", "scroffulous",
   "foetid", "mildly plump", "clammy", "lame
   brained", "boring", "cheezy"]
insulting_nouns = ["complete wheelbarrow",
   "airhead", "nitwit", "sandled sock-wearer",
   "sheep stalker", "emotional basket case"]
smelly_things = ["garlic", "camembert", "arm
   pits", "my great grandma"]
bad_property = ["a predisposition towards
   counting door handles", "gigantic knees",
   "hairy ears", "a noisy tummy", "a toxic
   burp", "a wet handshake", "silly legs"]
```

This program will need some lists of strings that it can use to build up rude sentences. These are just a selection. If you like you can provide your own to make the insults somewhat ruder.

```
# Choose one of the items in a list at
   random
# things - a list of anything, e.g. a list of
   strings
def anyof(things):
    choice = randint(0, len(things)-1)
    return things[choice]
```

The first function is given a list. Any list can be indexed by any number between zero (the first item in the list) and one less than whatever its length is. It simply obtains a random number in this range and returns that element in the list, effectively choosing a random element in the list.

```
#  Take the lower case words that make a
#  sentence and make it look like a sentence:
#  starting with a capital letter and ending
#  with a 'period' (such as a full-stop or
#  exclamation mark)
#      words - a string of lower case characters
#      period - the character to put at the end
#          of a sentence
def sentence(words, period):
    return words[0].upper() + words[1:] + period
```

Indexing a string will refer to individual characters, so 'words[0].upper()' is the first character of the string held in the 'words' variable made upper-case (ie capitalised) using a string class' 'upper' method.

It's obvious what would be meant by 19 + 23 in Python, but what about adding two strings, such as 'forty'+'two'?

```
>>> 'forty'+'two'
'fortytwo'
```

Adding two strings results in another string which is made out of the characters from the first string followed by those of the second.

The final piece of information needed to understand how 'sentence' works is to understand the idea of a 'slice' of a list. A slice is a more complex way to index a list than giving a single cardinal number. In a slice, a starting and an ending cardinal number are given separated by a colon (:) and the result is a shorter version of the list, beginning at the first and ending at the item before the last. For example,

```
>>> ['a', 'b', 'c', 'd', 'e'][2:4]
['c', 'd']
```

(remember that the first item in a list is numbered 0).

If one of the pair is omitted it's understood to be the first item (if the first is omitted) or the last item (if the last is omitted). Thus the slice [1:] means all of the list from the second item onwards:

```
>>> "string"[1:]
'tring'
```

You should now be able to see that 'sentence' concatenates a capitalized version of the first character with the rest of the characters and adds the 'period' string to the end (which we expect to be either '.', '?' or '!').

```
# Prefix a noun by the correct indefinite
   article ('a' or 'an')
#      noun - a string starting with the name
       to prefix
def article(noun):
    if noun[0] in "aeiou":
        return "an "+noun
    else:
        return "a "+noun
```

The syntax 'some-list "in" some-other-list' evaluates to a Boolean value (true or false) depending on whether the first list occurs as part of the second, so 'noun[0] in "aeiou"' tests the first character of the value in the variable 'noun' to see whether it's either 'a', 'e', 'i', 'o' or 'u', so this function puts either 'a' or 'an' in front of a noun depending on whether it starts with a vowel.

```
# Make a long string by joining the items in a
# list of strings with one delimiter between
# the initial items and another between those
# and the final item.
#    things - a list of strings
#    delim - a string to put between the first
      strings in the list
#    lastdelim - a string to put between the
      last two strings
def concatenation(things, delim, lastdelim):
    if len(things) == 0:
        return ""
    elif len(things) == 1:
        return things[0]
    elif len(things) == 2:
        return lastdelim.join(things)
    else:
        return delim.join(things[:-1]) +
            lastdelim + things[-1]
```

The function 'len' always returns the number of elements in a list, so this function does something different depending on whether a list has none, one, two or more elements in it. The class for strings has an additional method called 'join'. This takes a list of strings and joins each of the strings together separating them by the value join is applied to. For example,

```
>>> "-".join(['a', 'b', 'c', 'd'])
'a-b-c-d'
```

When giving negative cardinal numbers in indices they refer to positions in the list relative to its end instead of its beginning, so:

```
>>> "string"[-1]
'g'
```

and

```
>>> "string"[:-1]
'strin'
```

This should give you enough information to explain the following:

```
>>> concatenation(["one"],"; ","," and ")
'one'
>>> concatenation(["one"], "; ", "; and, ")
'one'
>>> concatenation(["one","two"], "; ", "; and, ")
'one; and, two'
```

```
>>> concatenation(["one","two","three"], "; ",
    "; and, ")
'one; two; and, three'
>>> concatenation(["one","two","three","four"],
    "; ", "; and, ")
'one; two; three; and, four'
```

```
# Make a comma separated list of up to three
    randomly selected
# adjectives taken from a list of adjectives
#    adjectives - a list of strings (each an
        adjective)
def adjectival_phrase(adjectives):
    shuffle(adjectives)
    description = adjectives[0:randint(1,3)]
    return concatenation(description, ", ",
    " and ")
```

This takes a list of adjectives, shuffles them, takes the first one to three of them and joins the results with commas and "'and's".

```
# Make up a random insult based on one of a
    small number of templates
def insult():
    noun = anyof(insulting_nouns)
    person = anyof(["you", "your "+anyof
    (relatives+pets)])
    if person == "you":
        is_verb = "are"
        smells_verb = "smell"
        has_verb = "have"
    else:
        is_verb = "is"
        smells_verb = "smells"
        has_verb = "has"
    choice = randint(1, 4)
    if choice == 1:
        return person+" "+is_verb+"
        "+article(noun)
    elif choice == 2:
        return person+" "+is_verb+" " + \
            adjectival_phrase(bad_adjectives)
    elif choice == 3:
        return person+" "+smells_verb+" of " + \
            anyof(smelly_things+pets)
    elif choice == 4:
        return person+" "+has_verb+" " +
        anyof(bad_property)
    else:
        return "Ooops ... I can't think of
        anything"
```

Finally we start to put our insults together here. This function decides whether the subject of the insult is going to be in the second ('you') or third ('your dog') person and adjusts the verbs it'll use accordingly. The third-person subject will be a random choice of a relative or a pet. Then it makes a random choice between four alternatives:

- The first choice is to say that the subject is one of the insulting nouns.
- The second is to say that the subject is a list of bad adjectives.
- The third is to say that the subject smells of either a random smelly thing or a pet. ('smelly_things + pets' makes a list containing everything in 'smelly_things' first and then everything in 'pets'.)
- The fourth is to say that the subject has one of the bad properties.

So each of the strings returned from this function should insult either the reader or one of her relatives or pets.

```
# Produce something rude: either a simple
    adjective or from one to
# three insults joined together
def rudeness():
    choice = randint(1, 4)
    if choice == 4:
        return sentence(anyof(bad_
        adjectives),"!")
    else:
        insults = []
        for n in range(randint(1,3)):
            insults += [insult()]
        return sentence(concatenation(insults,
        ", ", " and "), ".")
```

The one thing that's better than a single insult is a lot of insults, so this function joins a few of them together. First it chooses a random number between one and four and chooses to return a simple adjective as an insult whenever the number is four (which ought to be 25% of the time). The idea here is not to produce this relatively simple rudeness too often. The other choice is made the remaining 75% of the time.

This starts with an empty list ('[]') and uses a 'for' loop to add in an extra insult on every iteration. The function 'range' returns a list containing each of the numbers from zero upwards to provide the given number of items. For example,

```
>>> range(3)
[0, 1, 2]
```

which has three consecutive integers in it.

The syntax Python requires in a 'for' loop needs us to specify a variable that will take on each of the values in this list, but we don't need to use it. The idea here is simply to execute the body of the for loop for as many times as the argument 'range' (which is a random number between one and three).

The special type of assignment '+=' means 'add the value on the right to the value on the left'. For example,

```
>>> l = ["one"]
>>> l += ["two"]
>>> l
['one', 'two']
```

So the for loop builds up a list of one to three insults. These are then joined together with commas and ands and finally properly capitalized as a sentence ending with a full stop (.).

```
# The main program
def main():
    print(rudeness())
```

It's a matter of good Python style to provide a function called 'main' that runs the whole program. In our case all it does is call 'rudeness' to return a string containing our insults and print it out.

```
if __name__ == '__main__':
    main()
```

Typing in these two lines completes the program. The whole thing can be saved as a file, perhaps called 'docker.py'. There are two ways that Python can typically use 'docker.py': it can either be run directly or it can be imported into another program that'll want to use the functions it defines itself. In the first case the special variable '__name__' will evaluate to the string '__main__' whereas it will evaluate to a different value if the file's been incorporated into another program. The effect of these final two lines, therefore, is to call 'main()' only if the file wasn't included by another Python program.

If you've used Geany you'll be able to execute the program by using the 'execute' button when it'll open a window containing the program's output.

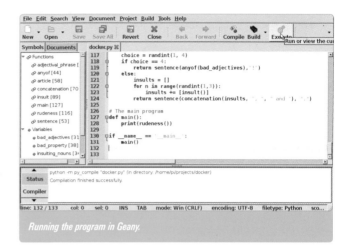

Running the program in Geany.

Other examples

You'll find further examples of Python programs elsewhere in this book. The program developed here is used as part of your own website in the section about providing a web server. A program to play the game of snake also has its own section, and a simple program that controls a toy can be found in the section about a 'twitter alert' project.

Help

Documentation about Python is available in several places, much of it on your Raspberry Pi.

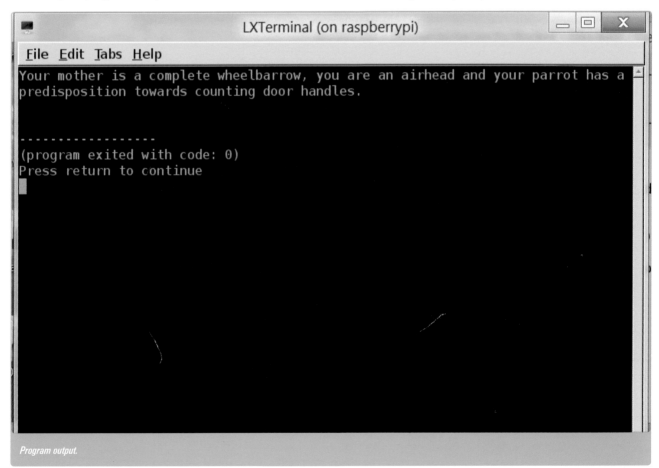

Program output.

Information about how Python and Python scripts can be run is contained in 'man',

```
man python
```

or 'info',

```
info python
```

Some frequently asked questions (FAQs) can be found in the file system in the HTML document

```
/usr/share/doc/python/FAQ.html
```

which can be opened in a local browser by using the URL

```
file:///usr/share/doc/python/FAQ.html
```

A lot of information is available inside Python using the 'help' function. This function takes a string containing information you'd like information about, but if you don't provide one it'll prompt you.

Any of the statements or functions available in Python can be investigated here. For example 'help ("if")' prints:

```
The "if" statement
********************

The "if" statement is used for conditional
execution:

    if_stmt ::= "if" expression ":" suite
              ( "elif" expression ":" suite )*
              ["else" ":" suite]

It selects exactly one of the suites by evaluating
the expressions one by one until one is found to
be true (see section *Boolean operations* for the
definition of true and false); then that suite is
executed (and no other part of the "if" statement
is executed or evaluated). If all expressions are
false, the suite of the "else" clause, if present,
is executed.
```

The most comprehensive source of information, though, can be found at the Python documentation website,

```
http://docs.python.org/
```

Python modules
One of the best things about programming in Python is that sometimes you don't have to. Python comes with a huge range of library modules as standard, just sitting there on your Raspberry Pi waiting for you to use them.

Using modules
We've already come across one of them above, in the 'random' module. This module provides quite a few functions including seed, getstate, setstate, jumpahead, getrandbits, randrange, randint, choice, shuffle, sample, random,

uniform, triangular, betavariate, expovariate, gammavariate, gauss, lognormalvariate, normalvariate, vonmisesvariate, paretovariate and webullvariate. If you'd like all of these function names to be available to your program you'd include this command at the head of your Python program:

```
from random import *
```

The provision of such a large number of names by a library isn't unusual and it may be that you'd like to consider the use of some of them as an error. For example, if you were only intending to use 'choice' and 'shuffle' you might use

```
from random import choice, shuffle
```

instead. This will also guarantee that your own use of a name such as 'uniform' isn't confused with its definition in the 'random' library.

An alternative is to leave the library's definitions in their own 'name space' so that every use must be preceded by the name of the module, meaning that the 'choice' function can be used by calling it 'random.choice'. This can be achieved by

```
import random
```

Modules already available
The full list of modules available can be listed interactively in Python using

```
help("modules")
```

Then information about all of the features exported by one of the modules, such as 'random', can be displayed using

```
help("random")
```

or typing 'random' into the 'help>' prompt. Finally, information about a specific name (such as 'random.choice') can also be requested using

```
help("random.choice")
```

Modules you can install
In addition to these standard Python modules, many more are available as Debian packages ready for you to install online using the command

```
sudo apt-get install <package>
```

To see a list of all the packages available use the command

```
aptitude search python2.7
```

This assumes that you're using Python version 2.7. You can check what version of Python you're using by running this command:

```
python --version
```

Until you know that you need any particular library it's sufficient for now simply to be impressed by the size of the list.

HOW TO COOK
Introduction To Linux

Although most desktop computers run a Microsoft Windows operating system there are many other devices around your home with operating systems of their own. Many of these are derived from the Unix operating system created at AT&T Bell Labs by Ken Thompson and Dennis Ritchie in 1970 (the first, little used, version of Windows was released 15 years later). The operating system you find in an Apple computer (MacOS), in an Apple iPod (iOS), in an Android phone (Android) or on your Raspberry Pi (Linux) are all in the same 'Unix' family tree. Today the fastest computers in the world predominantly run Linux.

Unlike many previous operating systems Unix was written in a high-level language, C, which had the great advantage that the whole operating system could be recompiled for different computers and so was 'portable'. This made it a popular choice for new computers, and Unix was reused, relicensed and rebranded several times. In 1986 the AT&T (by then 'system V release 2') design was well documented in the book *The Design of the Unix Operating System*, making copies inevitable. It was not, however 'free' in the sense described in the introduction, so when the University of California, Berkeley, modified and eventually completely replaced all its code and then made it freely available AT&T took the university to court.

Meanwhile, back at the Vrije Universiteit in Amsterdam, Andrew Tanenbaum (who's already been mentioned in the section about Python) needed a simple operating system to illustrate the ideas in his excellent book *Operating Systems: Design and Implementation*. The book included the free source code of 'MINIX' which, having the same kernel interface, could run the same applications that ran on an old version of AT&T Unix.

It seems that, influenced by MINIX's design, Swedish-speaking Danish student Linus Torvalds became terribly distracted while writing a terminal emulator for his IBM PC and accidentally wrote a MINIX-like operating system kernel instead. Wanting it to be a completely **free** 'Unix;' he initially called it 'Freakx', but it soon became better known as 'Linux' (pronounced as Lynn-ucks, not Lie-nucks).

Although influenced by MINIX, Andrew Tanenbaum didn't approve of Linux's design. Linus said from the very start of the project that pragmatic choices would be made to get it working as quickly as possible, to keep it simple and to make it powerful. More than 20 years later the kernel has passed through the hands of a worthy but small group of maintainers and still adheres to similar principles.

Linux distributions

A kernel alone is of very little use. When Linux was first announced there were no programs for it or the kernel to run. Even the compiler used to compile it, FSF's GNU C compiler, couldn't run under Linux and was available separately.

The attraction of adding, or perhaps even developing, new programs to add to this free core collection was that they would, in principle, allow a completely 'free' operating system to be built.

Since then the kernel has acted as a magnet for other free applications, all of which have now aggregated into huge collections called Linux 'distributions'.

Being free, anyone can collect programs together and make their own distribution. Typically a distribution will collect together an operating system that's tailored to a specific purpose, group of users, target computer or even operating system philosophy. Currently there are literally hundreds of them available. Three of the most prolific distributions (insofar as they've led to many related distributions) are Slackware, Debian and Red Hat. The reference distribution for Raspberry Pi is based on Debian.

The command-line shell

The original shell for the Unix operating system provided a command-line interface and was called 'sh'. It was replaced by a more advanced version by Stephen Bourne at AT&T in 1977, which, often referred to as the 'Bourne shell' to distinguish it from a rival 'C shell', was subsequently adopted in many operating systems based on Unix. While particularly good at supporting 'script' programs whose basic instructions were taken from the applications that the operating system provides, the Bourne shell wasn't quite as friendly as other shells in helping users as they typed new command-lines. Nor, crucially, was it 'free': it was a licensed part of the operating systems it appeared in.

In the early 1980s Richard Stallman initiated the 'GNU' project to create an entire version of Unix made only from free software. Although **G**NU's **n**ot **U**nix (hence its name), the one thing that it is is free. Richard Stallman even gave up his job at MIT to ensure that there'd be no possibility of their claiming ownership of the code his new Free Software Foundation would produce. Since then this body of commands has grown to incorporate virtually all the most important commands that were used in Unix.

The first version of Linux that Linus Torvalds announced in 1991, which could only be built on a MINIX system, contained only four files other than the kernel. One was the 'read-me' notes about the other three and another was the source program for 'bash' provided by Richard Stallman's Free Software Foundation (FSF). This is essentially a free version of 'sh'. Even today 'bash' is available in virtually every Linux distribution and is the program that provides the command-line interface in Raspberry Pi.

Our companion manual *LINUX Manual: Everything You Need to Get Started with Ubuntu Linux* also provides a great guide to Linux, especially if you want to use it as a desktop replacement for Windows.

Anatomy of a command line

Command descriptions
When describing commands there's a convention involving triangular brackets – less-than (<) and more-than (>) – being used, which prevents having to provide confusing examples. For example, if the 'cat' command has to be followed by the name of a file it's common to provide the following kind of description:

```
cat <file>
```

rather than providing an explicit example file name such as 'myfile':

```
cat myfile
```

The idea is that when you see '<file>' in the first description you imagine that it could be replaced by any file name. It helps to prevent the reader inferring, as she might from the second description, that 'cat' must always be accompanied by the text 'myfile'.

Similarly, several commands can take a series of arguments separated from each other by spaces. Rather than trying to give examples with different numbers of arguments we'll use an ellipsis (…) after the description of the thing that's to be repeated. For example, if the 'cat' command could be followed by any number of file names (which it can) it might be described using

```
cat <file>...
```

rather than providing examples

```
cat myfile
cat myfile anotherfile yetanotherfile
```

Sometimes a command line might or might not include something. To indicate things like this we'll use square brackets ([and]). For example, a 'cat' command can be followed by '-n' or it might not, but if '-n' does appear it must occur just after the 'cat'; so we might use the description

```
cat [-n] <file>...
```

rather than some examples:

```
cat myfile anotherfile
cat -n myfile anotherfile
```

This convention is used by a lot of the documentation about commands that's provided in Linux itself.

Typical command line
Most commands look like this:

```
<command> [<option>...] [<argument>...]
```

That is, the command name is optionally followed by any number of options and then any number of arguments (all separated by at least one space). The difference between an option and an argument is just that an option always begins with the minus character (-). The order of the arguments is usually important, but very often the order of the options is irrelevant.

Often, but not always, there'll be two versions of an option, one consisting of a single character (such as '-n') which is quick to type, and an equivalent beginning with two hyphens (--) which is more descriptive (such as '--number').

Some options are meaningful on their own, while others introduce the next thing on the command line which is considered to be part of the command line (for example, '-n 42').

The '<command>' itself might be built into 'bash' or might be the name of a file in the filing system to execute. Bash knows about a small number of directories in the file system where it'll look for the <command>. This means, for example, that because one of these directories is '/bin' typing the command 'cat' will use the file which is usually found at '/bin/cat'. This set of directories can be changed by the user and is called the 'path' that the command is looked up in.

Input and output
The shell provides every command with a single input from which it can read and two outputs to which it can write. One output is intended to be used for 'normal' output and the other only for error messages.

The shell can redirect both of these data streams in various ways, but by default the input stream comes from the terminal and the outputs are displayed on the terminal.

When a command reads its input from the terminal it'll continue to consume characters typed until a new line is started with Ctrl-D (a character that can be generated on a keyboard by pressing the 'Control' and the 'D' keys at the same time).

Help
This book doesn't aim to describe the options of each command available in Linux. In fact, it doesn't even aim to describe all the commands, just some common ones. Luckily there are ways to discover this information from Linux itself.

The first way to discover what a given command does is simply to ask it. Almost all commands respond to the '--help', or occasionally just the '-h' option to print out a brief summary of what they expect by way of options and arguments. For example:

```
pi@raspberrypi ~ $ cat --help
Usage: cat [OPTION]... [FILE]...
Concatenate FILE(s), or standard input, to
standard output.

  -A, --show-all         equivalent to -vET
  -b, --number-nonblank  number nonempty
                         output lines,
                         overrides -n
  -e                     equivalent to -vE
```

```
-E, --show-ends        display $ at end of
                       each line
-n, --number           number all output lines
-s, --squeeze-blank    suppress repeated
                       empty output lines
-t                     equivalent to -vT
-T, --show-tabs        display TAB
                       characters as ^I
-u                     (ignored)
-v, --show-nonprinting use ^ and M- notation,
                       except for LFD and TAB
    --help      display this help and
                exit
    --version   output version information
                and exit
```

With no FILE, or when FILE is -, read standard input.

```
Examples:
  cat f - g  Output f's contents, then
standard input, then g's contents.
  cat        Copy standard input to standard
output.

Report cat bugs to bug-coreutils@gnu.org
GNU coreutils home page:
  <http://www.gnu.org/software/coreutils/>
General help using GNU software:
  <http://www.gnu.org/gethelp/>
For complete documentation, run:
  info coreutils 'cat invocation'
```

A second method involves the 'man' command (short for 'manual'):

```
man <command>
```

This will normally provide a little more information than asking the command using '--help'. Many commands also have more comprehensive documentation that's displayed using the command:

```
info <command>
```

The 'info' command will start by presenting information about its argument, but also has a menu system that allows other topics and commands to be investigated.

Still more information (not all of it documentation) is available about many commands, or about the subsystem or package they're part of, in the directories held at

```
/usr/share/doc
```

Many of these directories are provided in HTML, so should be viewed using a browser. For example, there are some frequently asked questions about programming in the Python language held in

```
/usr/share/doc./python/faq/programming.html
```

These could be viewed in the Raspberry Pi GUI by typing the command

```
midori file:///usr/share/doc/python/faq/
programming.html
```

On the standard Raspberry Pi distribution there's a link to some of these local files about using Debian Linux command on the desktop, called 'Debian Reference'. Clicking this will open the web browser to inspect them.

If you have access to the Internet, typing 'linux <command>' into your favourite search engine is quite likely to provide useful information.

Naturally, the same search engine is also an excellent tool if you know what you want to do, but don't know which Linux command might do it. Linux can do a quick search of all of its manual pages for you to suggest commands that might be relevant too. To use this feature type

```
man -k <word>
```

where '<word>' is a word you hope to find in the one-line description held about each Linux command.

Directories of files
As discussed in the section about operating systems, a file name in a Linux file system begins with a slash (/) and continues with the names of directories, subdirectories, sub-subdirectories and so on each followed by a slash (/), finally ending in the name of a file in a directory. Such names can quickly become long and unwieldy so the shell remembers a working directory and allows file names to be used relative to that position.

The working directory
When a working directory is set file names that don't begin with '/' can be used to refer to directories and files in it. Thus if the current working directory were

```
/home/pi
```

the name 'projects' refers to the same thing as

```
/home/pi/projects
```

and, if that's a directory containing another object called 'docker', that object could be referred to either as

```
projects/docker
```

or

```
/home/pi/projects/docker
```

The parent directory name
Locations 'above' the current working directory can also be reached, because every directory contains a subdirectory called '..' which is actually the directory

that contains it. So, in our example above the directory '..' would actually refer to

```
/home
```

and '../..' would refer to the root directory

```
/
```

Given that /home contains directories for each user, if you've created a new user called 'mum' she'll have a directory that you could refer to as

```
../mum
```

User directories
Accessing other users' directories is so common that the shell has a special way to refer to them, by starting the file name with a tilde (~) and the user's name, so that the file 'cake-recipe' in mum's directory can be named

```
~mum/cake-recipe
```

The tilde character on its own refers to the logged-on user's own directory, giving yet another way to refer to the 'projects' directory:

```
~/projects
```

```
cd <directory>
```

will change the current working directory to <directory> (cd is short for **c**hange **d**irectory).

```
pwd
```

prints the current **w**orking **d**irectory (in case you've forgotten where you are).

```
pushd <directory>
```

(**push d**irectory) remembers the working directory before changing it to <directory>.

```
popd
```

(**pop d**irectory) returns the working directory to the last one remembered by 'pushd'. Working directories remembered by previous 'pushd's will be selected by subsequent 'popd's.

```
ls [<directory>]
```

lists the files in <directory>. If no <directory> is given the files in the current working directory are listed.

```
cp <fromfile> <tofile>
```

makes a **c**opy of <fromfile> and calls it <tofile>.

```
cp <from>... <todirectory>
```

makes a **c**opy of each <from> with the same name but inside the directory <todirectory>.

```
mv <fromfile> <tofile>
```

moves the file <fromfile> so that its new name is <tofile>. If possible this is done simply by renaming the file, but if this is impossible a copy of <fromfile> is made called <tofile> and then <fromfile> is deleted (which can be a lot slower than renaming for large files).

```
mv <from>... <todirectory>
```

moves each <from> with the same name but located inside the directory <todirectory>.

```
rm <file>...
```

removes the given <file>s from the file system. Once removed the contents of the files cannot be recovered.

```
ln -s <from> <to>
```

creates a **lin**k of a ('symbolic') file name <from> in file <to>. Following this <to> behaves almost exactly as if it were <from> and essentially becomes another name for it.
   The command supports other types of similar links, but symbolic links have the great advantage that they're easily visible when they're listed using 'ls'.
   This command cannot be used to create symbolic links in file systems that don't support them. For example, the Microsoft file system used for the boot partition at /boot can't have links created in it.

```
mkdir <directory>
```

makes a new **dir**ectory called <directory>.

```
rmdir <dir>...
```

removes the given directories from the file system. Once removed the directories cannot be recovered. However, only empty directories can be removed.

```
rm -r <directory>...
```

removes the given <directories>s from the file system, entering the directory to delete the files in it and recursively entering subdirectories to do the same. The contents of none of the removed files can be recovered.

```
touch <file>
```

creates a new file called <file> if it doesn't exist, otherwise updates its last modification time to 'now' without modifying the file.

Security

Not everyone is allowed to read, write and run (execute) every file. Instead, each file is associated with some security information dictating which of these types of access are allowed to three groups of subjects: the file's owner, people in the file's group, and everyone else. Both the file's owner and its group are also associated with the file as numeric identifiers.

The security information associated with a file (or directory) can be displayed using

```
ls -l <file>
```

for example:

```
pi@raspberrypi ~ $ ls -l test.py
-rwxr-xr-- 1 pi users 518 Aug  5 16:59 test.py
```

In this listing the security information is represented as ten characters '-rwxr-xr--', its owner is 'pi' and its group is 'users'. The 'ls' command translates the numeric identities of the user and group into a more legible form automatically. If required the actual numbers stored can be displayed like this:

```
pi@raspberrypi ~ $ ls -ln test.py
-rwxr-xr-- 1 1000 100 518 Aug  5 16:59 test.py
```

(user 1000 is 'pi', and group 100 is 'users').

The ten characters are notionally split into an initial character and then three groups of three characters. The first group of three characters ('rwx') shows the access that the user ('pi') has to the file (read, write and execute), the second ('r-x') shows the access that others in the file's group ('users') will have (read, no write, and execute), and the third ('r--') shows the access that everyone else has (read, no write and no execute).

The same controls are present on directories, but in this case the 'execute' permission is optionally replaced by an 'examine' permission that's needed to see the names of the files it contains (although this isn't used on Raspberry Pi). You must have 'write' permission to a directory to be able to create, delete or update any of the files in it. If you don't have 'read' access to a directory you won't be able to read the names of the files it contains.

```
whoami
```

will tell you which person is logged in and

```
groups
```

displays a list of the groups of which the logged-on user is a member.

```
umask <permissions>
```

Sets the permissions given to new files (the **u**ser mode **mask**). The permission can be expressed in the form 'u=<user permissions>,g=<group permissions>,o=<others permissions>', so that the example permissions above '-rwxr-xr--' would be set using 'u=rwx,g=rx,o=r'.

```
umask —S
```

shows the permissions that are given to new files as they're currently set.

```
chmod <permissions> <file>...
```

changes the file mode of each <file> according to <permissions>. In this case <permissions> can include changes to the permissions that already exist, so not all of 'u=', 'g=' and 'o=' need be specified, and the '=' may be replaced by '-' if a permission is to be removed or a '+' if it's to be added. For example, a <permission> of 'g-wx' will leave the file permission unchanged but remove any write or execute access that those in the file's group may have.

```
chgrp <group> <file>...
```

changes each <file>'s group to the one with symbolic name <group>. On Raspberry Pi the names of all the groups (and their corresponding numeric identifiers) are held in the file /user/group.

```
su [<user>]
```

After prompting for their password **s**ets the current **u**ser to the one with symbolic name <user>. Subsequent commands will be executed as user <user>. On Raspberry Pi the names of all the users (and their corresponding numeric identifiers) are held in the file /usr/passwd. If no <user> is given, set the current user to the privileged user 'root'.

```
sudo <command>
```

Having **s**et the **u**ser to root **do** the command <command>. It isn't necessary to provide 'root's password; instead the current user has to be allowed to use 'sudo' (as dictated by the content of the file /etc/sudoers) and must then normally give their own password. This can provide much more security than allowing many users to know 'root's password. In some systems this mechanism is relied upon so extensively that 'root' isn't given a password at all.

In Raspberry Pi /etc/sudoers allows only the main user 'pi' to use this command (although that user can use the 'visudo' command to update /etc/sudoers and alter this situation).

A number of commands can only be executed as the privileged user 'root', and this fact is indicated in the command explanations below by showing them run by 'sudo'.

```
sudo groupadd <group>
```

adds a new numeric **group** identifier and allocates it the name <group>.

```
sudo groupdel <group>
```

deletes the given **group** from the system.

```
sudo useradd <user>
```

adds a new numeric **user** identifier and allocates it the name <user>.

```
sudo userdel <user>
```

deletes the given **user** from the system.

```
passwd
```

change the **passw**ord for the logged-on user. This command prompts for the existing password first.

Files

```
cat <file>...
```

con**cat**enate the files together and write them to the output.

Tip

If you want to see the content of a small file, 'cat <file>' is a quick way to do it.

```
less <file>
```

print the file to the terminal one page at a time. (The original Unix command is called 'more'; 'less' has an extended set of features and, in particular, allows you to use the 'page up' key on the keyboard to go backwards towards the beginning of the file.)

```
find [<directory>] -name "<pattern>"
```

find and print the name of a file whose name matches the regular expression <pattern> in the given directory or its subdirectories. If a <directory> isn't given, search the current directory. The pattern can contain one or more asterisks (*) to indicate that any number of characters could be in that position. (There are, in fact, a large number of options that allow you to determine exactly which files are searched when a match is found, and what to do when that happens.)

```
grep "<pattern>" [<file>...]
```

find lines in a file **g**lobally matching a **re**gular ex**p**ression <pattern> and print those lines. As we've mentioned before, a regular expression can involve characters such as an asterisk (*) to match any number of characters on the line. If no file is given perform the search on the lines from the input.

```
head [-n <number>] [<file>]
```

write the **head** of the file (the first <number> lines) to the output. Write ten lines if the -n option isn't given. If no file is given take the first lines from the input.

```
tail [-n <number>] [<file>]
```

write the **tail** of the file (the last <number> lines) to the output. Write ten lines if the -n option isn't given. If no file is given take the last lines from the input.

```
tail -f [<file>]
```

write the **tail** of the file and then follow it. 'Following it' means that instead of finishing when the last line's been output, keep waiting in case any additional lines are added to the end of the file. If they are, print them out as they occur.

Tip

Many programs that run providing some service or another write their output to files in the /var/log directory. Using 'sudo tail -f <logfile>' will provide a nearly real-time copy of what's being written to the logs as it occurs.

```
tee [-a] <file>...
```

in addition to copying all input to the output send it to each of the <file>s, forming a kind of **tee**-junction in the data stream. If the -a option is given append the text to the end of the file if it already exists.

Tip

A number of files in /sys and /etc can be written only by root. The 'tee' command provides an easy way to write a single line to these with command lines such as: 'echo Last line | sudo tee -a <file>'.

```
wc [<file>]
```

perform a **w**ord **c**ount of the characters in each of the files. This command also counts the number of characters and the number of lines in each file. If no file is given count words taken from the input.

```
diff <file1> <file2>
```

compare <file1> with <file2> and print out the lines that are **diff**erent and lines that appear to have been added or deleted to <file1> in <file2>. This command produces no output at all if the two files are the same.

Processes

After having launched a command from the shell you'd normally wait for it to complete. It is possible, however, to return to the shell immediately by typing Ctrl-Z (that is, depressing the control and the Z key on the keyboard at the same time). This doesn't abandon the command, it merely 'stops' it. In this state it can be started again by typing the short command

```
%
```

However, if the command isn't started again you're free to initiate and possibly stop another job.

```
jobs
```

displays a numbered list of the **jobs** that you've started but stopped before they completed.

```
bg
```

starts the last stopped job again, but in the '**b**ack**g**round'. The command it restarts will run without your being able to interrupt it again with Ctrl-Z, and it will share your console so that, for example, any output it produces may be interleaved with output that other commands you type may generate.

Each of these jobs, which might all be running at the same time, run in their own process (as described in the section about operating systems).

 Tip

If you run a GUI program, such as the web browser 'midori', it continues to execute for as long as it's displayed, meaning that you can no longer use the command-line interface. To get the prompt back type Ctrl-Z and then use the 'bg' command to start midori running again.

```
ps
```

provides the **p**rocesses **s**tatus for all the processes that have been started on your behalf and

```
ps -axw
```

will produce a numbered list of all the processes that exist in the whole system, even those that you don't own. The number printed in the first column of this list is the Process IDentifier (PID) for the process.

```
kill <pid>
```

will make a request of the process with PID <pid> to terminate. This command will normally fail if you aren't the owner of the process. Running the command using 'sudo' ignores this ownership requirement, however. Not all processes are willing to terminate when requested.

```
kill -KILL <pid>
```

doesn't allow the process to be involved in the decision and terminates it forcibly.

It's possible for your Raspberry Pi to be used by more than one user at a time. One might be using a USB keyboard while another uses the serial line (which is described in a different section). If you've enabled it there are various ways for people using the same network to also log on.

```
w
```

displays **w**ho is logged on, what device they're using and what they're currently running.

Time
On its own Raspberry Pi doesn't know what the time is, but it's capable of keeping track of the time once it's known. When it's connected to the Internet Raspberry Pi will obtain the correct time using the Network Time Protocol (NTP).

```
date
```

displays the time and **date** Raspberry Pi believes is current.

```
sleep <number>
```

waits <number> seconds before the command completes.

```
uptime
```

among other information displays the amount of time that's passed since the computer was last rebooted.

```
watch <command>
```

executes <command> repeatedly (every two seconds) and each time highlights any of the output text that differs from the previous run.

Storage devices
The section about operating systems described the way that storage devices, such as SD cards, USB thumb drives or USB hard drives, can be divided into sections called 'partitions', some of which may have file systems in them.

Linux keeps track of the partitions that it knows about in a kernel-supported file at

```
/proc/partitions
```

In a system with only a typical SD card it'll have something like this in it:

```
pi@raspberrypi ~ $ cat /proc/partitions
major minor  #blocks  name

  179        0   7761920  mmcblk0
  179        1     57344  mmcblk0p1
  179        2   7700480  mmcblk0p2
```

For each of the lines in this file Linux will have made a device to represent a storage area. In Linux a device can be placed anywhere in a file system as if it were a file. Normally device files are placed automatically in the /dev directory. Thus from the contents of /proc/partitions we can expect the following files to exist:

```
/dev/mmcblk0
/dev/mmcblk0p1
/dev/mmcblk0p2
```

On Raspberry Pi /dev/mmcblk0 represents the whole of the SD card, and /dev/mmcblk0 and /dev/mmcblk0 represent two partitions on it. This is how 'gparted' displays the situation (this command isn't installed by default, but is easy to obtain using 'sudo apt-get install gparted'):

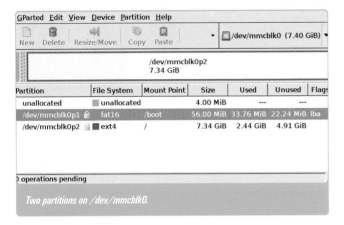

Two partitions on /dev/mmcblk0.

The bar shows two boxes, a very narrow (small) box for the first partition /dev/mmcblk0p1 and a much larger blue-grey one for partition /dev/mmcblk0p2. The whole thing, including these two partitions and the empty space around them, is /dev/mmcblk0.

Although it isn't always the case, both of these partitions are mounted into the file system: /dev/mmcblk0p1 is mounted on the root directory / so almost all Raspberry Pi's files are held on that partition; /dev/mmcblk0p2 is mounted at /boot.

```
sudo mount <device> <directory>
```

makes the existing <directory> the root of the file system held on the given <device>. Normally an empty <directory> is chosen because none of the files that it contains are visible once the <device> has been mounted on it. Instead all the files on the file system in <device> become visible there.

```
sudo umount <directory>
```

removes (**unmount**s) any device mounted at <directory>. Its previous content becomes visible again.

Although only available to the privileged user 'root' these devices can be read and written to as if they were normal files. Writing to an area of a device can destroy the filing system stored on it, so is generally a really bad idea.

```
sudo dd bs=1M if=<device> of=-<file>
```

will copy all of the data from the <device> to the <file> in blocks of 1Mbyte at a time (the copy may go faster when given larger block size). Typically partitions are huge (since they contain all the files in a file system), so you should take care that the partition that <file> is on has enough room.

```
sudo dd bs=1M if=<file> of=-<device>
```

will copy all the data in <file> on to the <device>. This should never be done if <device> is being used as part of the file system (ie mounted). Even if <device> isn't being used this command is likely to eliminate all previous data on the device.

```
du [<directory>]
```

shows how much **d**isk is **u**sed by all the subdirectories and files in <directory>. If <directory> isn't given, the current working directory is used instead. This command can take quite a long time to run.

```
df [<directory>]
```

shows how much of the **d**isk is **f**ree on the device that <directory> is part of. If <directory> isn't given, this command shows the space available on all devices that have file systems mounted on them. This includes artificial file systems that don't really have their own permanent storage locations.

```
sync
```

will make sure that any pending changes to any mounted device are made immediately, so that the actual state of the storage area is '**sync**hronous' with the intended state.

In principle it can be unsafe to take the power away from Raspberry Pi because file system updates to the storage area device aren't complete, which may mean that part of the file system wouldn't make sense if it were used as it is. In fact, most of the Linux file systems you might use have cleaver mechanisms in them to ensure that a senseless file system can be made self-consistent again when they're first mounted.

However, using 'sync' (and waiting for it to finish) before removing power can reduce the risk of file system corruption.

Software
Because Raspberry Pi runs Debian Linux it uses the Debian package system. New programs are distributed in packages, which are files the last part of whose name is '.deb'. The easiest way to obtain them is to use one or more 'repositories' of these files on the Internet. There are standard locations which always have the most up-to-date and suitable packages for your version of Debian. The names of the repositories that your computer uses are stored in this file:

```
/etc/apt/sources.list
```

If you 'cat' this file you'll probably find the name of a location that's devoted to Raspberry Pi packages. These packages are replaced almost continuously with improved versions, and new packages are added every now and then too.

You can instruct the package manager to find out what changes there have been in your repositories using the command

```
sudo apt-get update
```

It's then possible to request that all out-of-date packages are replaced by newer versions with

```
sudo apt-get upgrade

aptitude search <package>
```

will print a list of all the packages in your repositories whose name includes <package>. It isn't always simple to derive the name of a package that supplies a program you're interested in, but this command helps.

```
sudo apt-get install <package>
```

will find <package> in the Internet repository and install it on the computer. In order to do this it may also download and install any number of other packages that are required by the requested <package>.

 Tip

If you need to find the name of a package that provides a given file (for example a command) you can use the 'apt-file' command. Unfortunately this command may not be installed itself. You can install it using

```
sudo apt-get install apt-file
```

Before it can be used for the first time it needs to refresh the information it uses to search through, using

```
apt-file update
```

You can then search for a package using

```
apt-file search <file>
```

Output

```
echo "<text>"
```

simply writes <text> to the output.

```
printf "<formattext>" <argument>...
```

also writes <formattext> to the output, but can include escape sequences beginning with a back slash (\) and format specifications beginning with a per cent (%).

Escape sequences allow special characters to be included in the output. Perhaps the most useful of these is '\n' which will insert a 'newline' character that makes subsequent characters appear on a new line. Unlike 'echo' a newline character isn't automatically added at the end of the <formattext>, so several different 'printf's can be used to build up a whole line if required.

Format specifications each use up the next <argument> and define how they'll be displayed. For example, '%10s' will write <argument> out as a string padded out to a width of at least ten characters. For detailed information about format strings you'll need the manual page for the C function 'printf', which you may not have installed. To install it use

```
sudo apt-get install manpages-dev
```

then display the man page (from section three) using

```
man 3 printf
```

System errors

Many processes run at the same time, many of them providing or consuming services on the network. Very many of them provide logging messages about errors, warnings and other information. If all these messages appeared at the terminal it would be very distracting to the user, so they're generally written to log files. The standard directory in which to place log files is

```
/var/log
```

Because they might include sensitive information, such as passwords, most of these log files can only be read by the privileged user 'root'.

When something isn't working properly it's very likely that an error message will have been written to one of these files.

 Tip

```
sudo ls -ltr /var/log/
```

will list all the log files in /var/log ending with those most recently written to. These are often a good bet for looking at.

```
/var/log/messages
```

```
dmesg
```

will print the most recent lines that were written to the standard log output.

Environment

The shell maintains a small dictionary of 'environment

variables' which can be given textual values. Every program that runs has access to this dictionary and may change its behaviour depending on the value of specific variables. For example, 'bash' uses the value given to the environment variable 'PS1' to prompt the user.

```
<name>="<value>"
```

will set the shell's environment variable <name> to the given <value>. For example:

```
pi@raspberrypi ~ $ PS1="Oooh, you are
lovely..$ "
Oooh, you are lovely..$
```

Shell environment variables can be 'expanded' as part of a command line by preceding their name with a dollar ($). That is, before the command line is executed any occurrences of '$<name>' are replaced by the text value of <name> in the environment dictionary. For example:

```
pi@raspberrypi ~ $ animal="wombat"
pi@raspberrypi ~ $ echo "You are a $animal"
You are a wombat

env
```

prints the whole list of environment variables and the values they have set.

Although these variables are available in the shell they don't automatically get inherited by programs that are run. For other programs to 'see' the value of a local environment variable it must be exported.

```
export <name>
```

makes the local environment variable part of the environment that's given to all programs that are run under the shell.

Tip

The easiest way to see what value an environment variable has is simply to type

```
echo $<name>
```

Make your own commands

```
alias <commandname>="<text>"
```

will use <text> in place of the command name <commandname>. The upshot of this is that you can provide your own names for commands that you commonly use. For example:

```
pi@raspberrypi ~ $ alias packages="aptitude
search"
pi@raspberrypi ~ $ packages xara
p   xara-gtk          - GTK+ utility for
                        searching the Deb
v   xara-gtk-byte                         -
```

Tip

If you've written your own commands you may want to create a directory where you put them all and then update the PATH environment variable to include that directory. This will mean that your new commands will be run when their name is typed. For example, if you use the 'bin' subdirectory of your home directory you could add it to your path using

```
export PATH="$PATH:$HOME/bin"
```

There are many environment variables with specific uses. Here are a few:

Variable	Typical value
Description	
USER	pi
The name of the current user.	
HOME	/home/pi
The home directory for the current user.	
DISPLAY	localhost:11.0
The display that programs will use when they use a graphical user interface.	
PATH	/usr/local/sbin:/usr/local/bin:/usr/sbin:/usr/bin:/sbin:/bin
This is the 'path' we described above containing the names of a series of directories separated by colons (:). If the first item on a command line is the name of an executable file in one of these directories it will be run.	

```
unalias <commandname>
```

will remove an existing alias and

```
alias
```

will list the aliases you've already created.

A similar effect to an alias can be obtained by defining a simple shell function:

```
name() { <command> "$@"; }
```

For example:

```
pi@raspberrypi ~ $ packages() { aptitude
search "$@"; }
pi@raspberrypi ~ $ packages xara
i   xara-gtk          - GTK+ utility for
searching the Deb
v   xara-gtk-byte     -
```

(If you try these examples you'll need to unalias 'package' first.) The advantage of shell functions is that they can include more than one command and can, in fact, be rather complicated.

```
source <file>
```

will take commands directly from <file> instead of you having to type them in yourself. This can be useful, for example, if <file> contains a lot of function and alias definitions for you to use later.

If you have a file with the name '.bashrc' in your home directory Bash automatically sources it for you when it's run. In there you can place functions and aliases that you always want to be available.

It's also possible to create a file which executes any interpreted text simply by putting the program text into a file preceded by a line starting

```
#!<interpreter>
```

and making sure that the file is executable using

```
chmod +x <file>
```

The <interpreter> on the first line must be the full name of the file containing the interpreter program. Both Python and Bash are examples of interpreters. Bash is always available in the file /bin/bash so an example of a Bash script might be

```
#!/bin/bash
echo "Hello World"
```

If this is placed into a file on your path (discussed above) then typing its name will execute it using Bash. Here's a full example using several things we've described for you to follow through:

```
pi@raspberrypi ~ $ mkdir cmd
pi@raspberrypi ~ $ printf '#!/bin/bash\necho
  "Hello World"\n' > cmd/hello
pi@raspberrypi ~ $ cat cmd/hello
#!/bin/bash
echo "Hello World"
pi@raspberrypi ~ $ chmod +x cmd/hello
pi@raspberrypi ~ $ export PATH="$PATH:$HOME/
  cmd"
pi@raspberrypi ~ $ hello
Hello World
```

```
type <command>
```

will tell you what kind of thing <command> is and something about its definition. For example:

```
pi@raspberrypi ~ $ type packages
packages is aliased to `aptitude search'
```

or

```
pi@raspberrypi ~ $ type packages
packages is a function
packages ()
{
    aptitude search "$@"
}
```

and

```
pi@raspberrypi ~ $ type hello
hello is hashed (/home/pi/cmd/hello)
```

Shutting down

It isn't advisable simply to unplug Raspberry Pi while it's active (which, because it runs lots of programs in the background, is all the time). In some cases it's possible to corrupt the file system by doing this. It's far better to close down the computer with a command and then take the power away.

```
sudo halt
```

will stop all the processes, make the file systems safe and then turn Raspberry Pi off safely.

```
sudo reboot
```

will do the same but then start executing again so that Raspberry Pi reboots.

```
shutdown -r <time> <message>
```

will also reboot the system, but at the given time (which might be 'now', for example), and only after issuing the explanatory <message> on all the user's terminals.

Start-up and booting

The first program that the kernel loads from the file system is /sbin/init. On Raspberry Pi this program reads a configuration file '/etc/inittab' which provides details of how to start and stop the system, and how to set up terminal streams for users to log on through. Part of the start-up detail is selecting which of a small number of 'run levels' in which to start the system up. Not all of these run levels are useful but the following may well be distinguished:

Run Level	Detail
1	Stand-alone single-user: relying on a minimum number of devices and not requiring the user to log on. No network is normally set up and only one user is expected. This run level is very useful if the system has developed a problem that prevents the use of more complex run levels.
2	Network-mode multi-user terminals: providing the ability for more than one user to log on via different terminals and consoles and supporting the network and providing network services. The computer must be used by typing textual commands; there are no windows or other graphical controls.
5	Network-mode multi-user graphical user interface: similar to run level 2 except the computer can be used via a graphical windowing environment.

The '/etc/inittab' file normally indicates which of these run levels the system is to use, although this can be overridden by options appearing on the kernel command line (which we describe in the annexe about configuration)'.

To bring Linux up in run level 2 each of the programs in the directory '/etc/rc2.d' are executed (they're usually Bash scripts). For run level 5 the program in '/etc/rc5.d' is used and so on. Over the past few years this area of a typical Linux distribution has been updated to add various innovations that help reduce the amount of time it takes for the system to become fully usable. Typically they involve trying to run these scripts all together at once rather than one-by-one. Normally each of these scripts will enable some feature or another and will print a line to the console indicating what it was and whether its initialization succeeded or failed.

It isn't until the last of these 'init' scripts has been run that the user is invited to log in. A list of most of the possible scripts can be found using

```
ls /etc/init.d
```

The current state of any of these, for example 'cron', can be found like this:

```
sudo service cron status
```

They can also normally be started and stopped, for example

```
sudo service cron stop
```

Stopping a service isn't permanent. If the service was one that's normally started in your run level it'll be started again when Linux next boots. You can remove a service (such as cron) altogether (from all run levels) using this command

```
sudo insserv -r cron
```

following which it won't be run when Linux reboots (which is a shame, because cron is really useful).

A service can be included again in all its default run levels using

```
sudo insserv cron
```

Networking

A later section about writing programs that use the Internet describes a little about the way most communication between computers uses Internet Protocol (IP) addresses. There are many ways to use Raspberry Pi from other computers, which we also discuss later in the book, and all of them will require an IP address.

Finding an IP address

Unless you're lucky enough to have a network of computers at home with its own name server, which you'll have set up to give a name to your new Raspberry Pi, there will be no convenient name to refer to it from other computers. To overcome this problem you'll have to use its Internet Protocol (IP) address.

Your home network contains a Dynamic Host Configuration Protocol (DHCP) server which gives Raspberry Pi an IP address every time it boots. Unfortunately you may find that this address is different on each boot (although you may not). If so you'll have to discover Raspberry Pi's IP address each time you turn it on before you use it remotely.

If you don't boot into the graphical interface and you watch carefully as Raspberry Pi boots, you'll see that it displays its IP address just before asking you to log on. If you miss it, however, you'll need to find it a different way.

The following command will display your IP address:

```
hostname -I
```

It'll normally look like four numbers lower than 256 separated by dots. If the first of the numbers is 127 it may be that you don't have an Ethernet or WiFi connection. It's quite likely (although not inevitable) that the first two numbers are 192 and 168 respectively.

Some further commands for investigating the network are presented in the section about using the Internet.

04.
Software recipes

SOFTWARE RECIPES
Running Programs Regularly

It's possible to run the same program on Linux regularly (for example every hour or at the same day every week or month). This provides the opportunity to summarize information from attached hardware, tidy up log files, fetch new data from the Web, update software, test the latest version of your fish tank environmental control program, and so on.

All Unix-style operating systems support a background process (often called 'daemons') called 'cron'. Cron was originally created by one of the original authors of Unix, Brian Kernigham, and is responsible for responding to a number of cron tables provided both by the system and by individual users that specify what should be run when.

Preparation
It's very unlikely that cron hasn't been installed on your Raspberry Pi. You should be able to verify that cron is running properly as follows:

```
pi@raspberrypi ~ $ service cron status
[ ok ] cron is running.
```

If, instead, the response mentions an 'unrecognized service' the following will install it (assuming Raspberry Pi has an Internet connection):

```
sudo apt-get install cron
```

Editor
Changing cron tables will involve the use of the 'default' text editor. If you haven't considered which editor to use now might be a good time to select one.

The nano editor provides helpful documentation.

The editors that are available to you, the one that's currently being used and an opportunity to change it to another one, is provided using

```
sudo update-alternatives --config editor
```

For example, if you've installed the package 'joe' and you prefer the editor 'jstar':

```
pi@raspberrypi ~ $ sudo update-alternatives
--config editor
There are 8 choices for the alternative editor
(providing /usr/bin/editor).
```

Selection	Path	Priority	Status
* 0	/usr/bin/joe	70	auto mode
1	/bin/ed	-100	manual mode
2	/bin/nano	40	manual mode
3	/usr/bin/jmacs	50	manual mode
4	/usr/bin/joe	70	manual mode
5	/usr/bin/jpico	50	manual mode
6	/usr/bin/jstar	50	manual mode
7	/usr/bin/rjoe	25	manual mode
8	/usr/bin/vim.tiny	10	manual mode

```
Press enter to keep the current choice[*], or
type selection number: 6
update-alternatives: using /usr/bin/jstar to
provide /usr/bin/editor (editor) in manual mode.
```

The text editor selected in this way will also be used by any other user Raspberry Pi might support. You'll be able to edit a file (say 'myfile') using the command

```
editor myfile
```

If you wish to set your own editor, separately from the system default editor, you can set the 'EDITOR' environment variable. Following the command

```
export EDITOR=/usr/bin/jstar
```

any update of the cron tables will use the 'jstar' text editor.

To make this selection 'permanent' add the above line to the file of Bash commands that's executed every time you log on (~/.bashrc). (Naturally, to perform that update you'll have to run the editor of your choice.)

As a quick guide to some command-line text editors available under Linux:

Editor	Notes
vi	This is the standard text editor used by system administrators. It's a nightmare for first-time users because it's 'modal', with single-letter commands meaning different things depending on which mode the editor's in. In other respects it's fast, comprehensive and always available.
nano	This is very simple to use and constantly displays the command keys that can be used along with what they do. It isn't as capable as other editors but is fast and is normally installed in Raspberry Pi.
ed	This is almost the last surviving 'line mode' editor that can be virtually guaranteed to be present on every Unix-style system. First written by Ken Thompson, one of the original Unix authors, it is the only editor in this list that doesn't display a whole page of text and provide a What You See Is What You Get (WYSIWYG) interface; instead it operates on a current line which it reprints only once you've executed a command that updates it. It's tiny, super-fast, and even less friendly to use than vi. Once upon a time all text editors were like this.
emacs	This is another standard editor used widely by Unix users. It's huge and incorporates its own programming language for configuring it. Consequently almost any task can be accomplished with emacs, including ones that have nothing to do with editing text files. It can take even longer than 'vi' to master properly but it's reasonably fast under Raspberry Pi and can also be used from the GUI. Normally this editor needs to be installed separately, using `    sudo apt-get install emacs`
jstar	I like this editor. It's small, simple and uses the well thought-out command keys that the ancient 'wordstar' editor used. This editor also needs to be installed separately, using `    sudo apt-get install joe`

Email

Believe it or not your Raspberry Pi has its own email system provided by a system called 'sendmail' for sending messages from one user to another. If you have mail you'll be told between command executions

```
You have mail in /var/mail/pi
pi@raspberrypi ~ $
```

Such messages can be read and sent using the 'mail' command.

You'll find that, by default, an error in a command executed as specified by a 'crontab' command is emailed to you in this way. It can be more convenient to have these results emailed instead to your own email address.

The 'sendmail' command can actually be provided by more than one package in Debian and you may find that your Raspberry Pi uses the 'exim4' implementation. You can verify this using the command

```
ls -l /usr/sbin/sendmail
```

which may end '-> exim4' indicating that Exim is being used.

In principle 'sendmail' allows your Raspberry Pi to be used as a full Mail Transfer Agent (MTA), sharing the responsibility with all the other MTAs on the Internet to forward and deliver people's email. In this very responsible role your Raspberry Pi must be properly identifiable and have its own name in the Internet's Dynamic Name Service (DNS), including a fully qualified domain name. It's by no means impossible to obtain your own domain name for your little computer either at no cost with some inconvenience or with some cost and less inconvenience. However, you may not consider this effort worthwhile if you don't need your machine to fulfil such a demanding role.

If you have an email program such as 'thunderbird', 'outlook' or Windows Mail you'll have had to supply it with details about your mail server, which is normally accessed using the Simple Mail Transfer Protocol (SMTP). These will include the Internet address of the computer that you send SMTP traffic to as well as a name and password to prove to it who you are (sometimes but not always).

In general you'll find that email sent using Exim's implementation of 'sendmail' will result in other MTAs complaining that Raspberry Pi doesn't have a 'FQDN' (fully qualified domain name). Debian provides a simplified version of 'sendmail' in the package 'ssmtp' which can use your mail server details on your behalf to send email. To install this, use

```
sudo apt-get install ssmtp
```

Note that because Debian knows that exim4 and ssmtp are different ways to provide the same functionality, this will also uninstall exim4.

Before this can be used to send email you must add your mail server details into its configuration file using

```
sudo editor /etc/ssmtp/ssmtp.conf
```

The Google mail SMTP server makes a good example because it requires a lot of options to specify. This is an example of this file, with lines that need adding for Gmail shown in bold.

```
# Config file for sSMTP sendmail
#
# The person who gets all mail for userids < 1000
# Make this empty to disable rewriting.
root=postmaster
# The place where the mail goes. The actual
#  machine name is required no MX records are
#  consulted. Commonly mailhosts are named
mail.domain.com
mailhub=smtp.gmail.com:587
# gmail uses Transport Layer Security (TLS)
UseSTARTTLS=YES
AuthUser=Me
AuthPass=mypassword

# Where will the mail seem to come from?
#rewriteDomain=

# The full hostname
hostname=raspberrypi
```

```
# Are users allowed to set their own From: address?
# YES - Allow the user to specify their own
  From: address
# NO - Use the system generated From: address
# FromLineOverride=YES
FromLineOverride=YES
```

Here we've used 'Me' and 'mypassword' in place of your own Gmail login details.

 Warning

If you've placed your password in this file you should be concerned about who can read it and how much access people from the Internet have to your Raspberry Pi. If you also use Google Checkout, for example, these details may be sufficient to allow real money to be stolen from you. So, for example, change your Raspberry Pi password (in fact consider becoming a new user), make sure sudo can't be used by anyone else, protect your Raspberry Pi, and don't give the SD card to anyone else. Create a new user name for visitors to use.

Other, more standard, SMTP servers don't use TLS and so won't require the line starting 'UseSTARTTLS'. Also you may notice that Gmail's SMTP computer name is followed by ':587', which is the port on the computer that listens to SMTP traffic. Other SMTP servers listen on port 25, which is assumed if the colon and number are omitted.

In addition, if you use the SMTP server provided to you by your Internet Service Provider (ISP) you may find that it authenticates you simply by knowing who you are in other ways (ISPs always know who you are, so that they can charge you every month). In that case you may find that the two lines beginning 'Auth' are also unnecessary.

Test your ability to send email using a command such as

```
echo "Yea, Raspberry Pi can send emails" |
mail -s testing me@gmail.com
```

(You'll need to replace 'me@gmail.com' with your own email address.) Check that the email arrives.

Cron tables

Cron uses one special table held in /etc/crontab to direct system-wide regular activities but otherwise uses cron tables that are kept for each user.

Except for comment lines each line of a user's cron table defines a regular time and a command to be executed. A comment line begins with a hash character (#) and is ignored. Other lines are broken up into five time-items separated by spaces and a command that takes up the rest of the line. The time items occur in the order

1 Minute (from 0 to 59).
2 Hour (from 0 to 23).
3 Day of the month (from 1 to 31).
4 Month number (from 1 to 12).

5 Day of the week (with 0 and 7 being Sunday, 1 being Monday etc).

Each can be either an environment variable setting or a comma-separated list of numbers that have to match the current time, or a single asterisk (*) character which indicates that the item doesn't matter.

Here are some example time specifications:

Time items	Meaning
0 * * * *	Hourly (exactly on the hour).
0 0 * * *	Daily (exactly at midnight).
0 4 * * 3	Every Wednesday (at 4am).
30 19 * * 6,7	At 7:30pm every weekend.
0,15,30,45 * * * *	Every quarter of an hour.
0 0 1 * *	Monthly (at midnight on the first of the month).

The rest of the line will be executed by the system shell as if the user had run the command. The system shell is normally '/bin/dash' in Raspberry Pi, but almost everything in the section about Bash applies to Dash.

The 'crontab' command is used to view and update cron tables. Initially users don't have one but a new one will be created when it's first edited using

```
crontab -e
```

This will enter your selected text editor with the contents of a new cron table ready for altering. On the first run the new contents of the file are just comment lines explaining the format of the file.

Cron will use the environment variable 'MAILTO' to determine where to send email and you can set this, for example to 'me@gmail.com', in your cron table by including a line that precedes the entries that give commands to be executed, such as

```
MAILTO=me@gmail.com
```

This will mean that any output your command makes will be emailed to this address. If your command doesn't create any output you'll be emailed only if it fails (that is, leaves a non-zero return code). For commands that are executed very often you may wish to wrap them in a Bash script that diverts all output to a log file to avoid excessive emails.

Help

The cron daemon is explained by its own man page:

```
man cron
```

as is the crontab command:

```
man crontab
```

while the detailed format of a cron table file is described by section 5 of the manual pages displayed using

```
man 5 crontab
```

SOFTWARE RECIPES
Scraping Web Pages

A web browser allows you to interact with any web server on the Internet and view pages. But what if your software needs the content of a web page for further manipulation?

Those who saw the film *The Social Network* will have seen a portrayal of Mark Zuckerberg sowing the seeds of his idea for Facebook in a late-night scripting session to collect photos of all the women at his university, using the 'wget' command. The 'wget' command allows web pages to be downloaded and stored for use in programs and scripts. It is, however, not the only command available to perform this kind of task, and today a newer command, 'curl', has become more capable.

How it works
In the section about providing a web server on Raspberry Pi we describe what a web address, or Universal Resource Locator (URL), means. The 'curl' command supports a large number of protocols used in URLs, including those for reading email information and information from SSH (a full list can be obtained using 'man curl'), but 'wget' supports only FTP, HTTP and HTTPS (which are perhaps the most useful).

Either of these commands can be used as a web client using an Internet connection to support the required protocol.

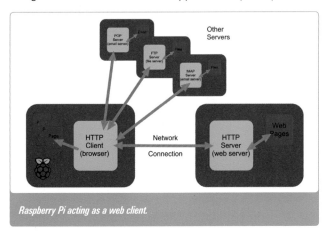

Raspberry Pi acting as a web client.

Preparation
It's quite likely that the commands 'wget' and 'curl' are already supported in your Raspberry Pi distribution. If they aren't they can be installed using the command

```
sudo apt-get install curl wget
```

Downloading web pages
The command

```
curl --output <file> <URL>
```

or

```
wget --output-document <file> <URL>
```

will fetch the web page named in the URL and store it in the given file.

Some web servers will only provide a page if they believe that the web client is advanced enough to interpret it, making a judgement based on the 'user agent' identification that the client provides. It's possible to set the user agent that'll be quoted by adding it to the command line before the <URL>:

```
--user-agent="Mozilla/5.0"
```

which will pretend to be a Firefox browser (which is almost always accepted).

Scraping web pages
The process of extracting information from an HTML web page using a script is called 'scraping'. Ideally you will have read about HTML in the section about building a web server before trying your own scrape. Complex web pages tend to be more difficult to scrape because they normally contain extraneous information that you need to try to avoid.

Manual extraction
If we know the structure of certain web page elements, such as images or URL links, we can easily extract them from the HTML file. For example, this command

```
cat mywebfile.html | sed 's.<.\n<.g' | \
    sed -n 's/[^<]*<img.*src="\([^"]*\).*/\1/pg'
```

uses the very flexible 'sed' stream editor command to find all the images in mywebfile.html and print out their URLs. (In detail this command puts every HTML element beginning '<' on a new line and then finds all the lines that begin '<img' and prints those lines, having removed everything except the text between quotes after 'src="'.)

Similarly, this command will extract all the links:

```
cat mywebfile.html | sed 's.<.\n<.g' | \
    sed -n 's/[^<]*<a.*href="\([^"]*\).*/\1/pg'
```

If you do this you may notice that these URLs are often relative to the page they were found in, so that if they don't begin with 'http:' (or 'https:', or 'ftp:' etc), the directory part of the URL where these pages are found is presumed. For example, if an image URL

```
images/raspberry.jpg
```

is found in a page

```
http://fruit.com/soft/pictures.html
```

the actual web page being referred to is

```
http://fruit.com/soft/images/raspberry.jpg
```

Precision extraction

Web pages consist of a hierarchy of parts with sub-parts inside them. The Firefox browser has an excellent way to demonstrate this three-dimensionally. Right-clicking on any item on a web page produces a menu including 'Inspect Element (Q)'. Using this will isolate the area of the screen covered by the HTML that displays it and will show all the HTML elements it's inside at the bottom of the page. There's a '3D View' button that'll make this hierarchy even more obvious.

XPaths

Along the bottom of the screen you will see the path including elements that surround the one that was selected for investigation. It shows that an <a> (which is an HTML hyperlink) was selected, which was inside a <div>, which was inside a <span>, which was inside a <div>, and so on.

It's possible to identify any element on a web page by knowing exactly what the path to it was. You'll see that some of the elements at the foot of the screen include a hash (#) or a full stop (.) and then a name after them. In Firefox you can see the whole of this name by right-clicking on the relevant part of the path. It will display the names of all the other elements contained in the same outer level too (its siblings). These names make it easier to find the particular example of the element among its siblings. A hash means that the element has an 'id' attribute with that name, and a full stop means it has a 'class' attribute with that name.

There's an entire standard language, called XPath, whose only purpose is to identify part of an HTML document using one of these paths. For example, the path to a <div> which has an 'id' attribute of 'softfruit' which occurs in the <body> of the <html> can be found using an XPath

```
/html/body/div[@id="softfruit"]
```

If a 'class' or 'id' attribute isn't available to distinguish an entry from its siblings you can simply identify the one that you want using a sibling number. For example, the third row of a table found inside the <div> above could be referred to as the third sibling <tr> element of the table:

```
/html/body/div[@id="softfruit"]/table/tr[3]
```

The information displayed along the bottom of the screen doesn't identify this index number but is certainly a help in finding the XPath that would select that part of the web page.

Extracting precisely the information you need from a web page is a first step to providing a Bash or Python script to use that information.

xmllint

The command 'xmllint' is an extremely useful tool in helping you investigate the structure of a web page. It can be installed on Raspberry Pi using

```
sudo apt-get install libxml2-utils
```

The Hyper-Text Markup Language (HTML) used for web pages is very nearly an example of a larger class of languages that can be created by the eXtensible Markup Language (XML). With the tiniest of tweaks a program that understands XML can also understand HTML.

As part of its main job of helping find suspicious or just bad XML, xmllint offers a command-line interface allowing the user to investigate the structure of XML (or HTML) a little as if it were a file system in Linux.

It uses many of the commands that were discussed in the section about Linux for navigating directories (such as cd, ls, and pwd) to navigate XML entities, using path names in place of directory names. In addition, an XPath can be used at any point to describe a selection of the file which can be then used to create a separate XML (or HTML) file.

Worked example

Suppose that as part of my system I want to find the time of the first bus every morning in order to create an alarm that's in-tune with the bus company's timetables.

With very little work (a web search) I can find a web page that provides this information at URL

```
http://www.travelineeastanglia.org.uk/ea/
XSLT_TTB_REQUEST?language=en&command=
direct&net=ea&line=20003&sup=B&project=y
08&outputFormat=0&itdLPxx_displayHeader=
false&itdLPxx_sessionID=EMEA_98_21772312&
lineVer=1&itdLPxx_spTr=1
```

which is a page like this:

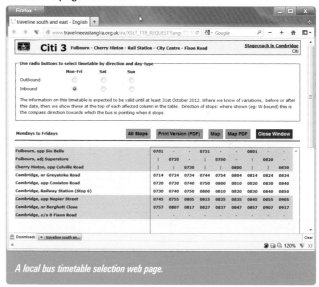

A local bus timetable selection web page.

I can put the HTML for this page in a file called timetable.html using the two commands

```
url="http://www.travelineeastanglia.org.uk/
    ea/XSLT_TTB_REQUEST?language=en&command=
    direct&net=ea&line=20003&sup=B&project=
    y08&outputFormat=0&itdLPxx_
displayHeader=false&itdLPxx_sessionID=
    EMEA_98_21772312&lineVer=1&itdLPxx_spTr=1"
    curl  --user-agent "Mozilla/5.0" --output
    timetable.html $url
```

which I can now explore using 'xmllint' as a shell:

```
pi@raspberrypi ~ $ xmllint --html timetable.
html --shell
/ > ls
timetable.html:80: HTML parser error :
Unexpected end tag : input
his.value)" id="weekdayH_1" tabindex="1"
checked="checked" value="xxx1H"></input
  ^
```

(The main task of xmllint is to spot poor HTML, and this page, like many others, has errors in it that xmllint finds and prints out. They're not severe enough to worry us, though.)

```
/ > cd html
html > pwd
/html
html > ls
---         10 head
-a-          5 body
html > cd body
body > ls
t--          2
---          3 center
t--          2
-a-          1 script
-a-          1 script
body > cd center
center > pwd
/html/body/center
center > ls
t--          2
-a-          2 div
t--          2
center > xpath *
Object is a Node Set :
Set contains 1 nodes:
1   ELEMENT div
    ATTRIBUTE style
      TEXT
        content=position:absolute;overflow:hidde
        n;text-al...
    ATTRIBUTE id
      TEXT
        content=htmlTT
center >
```

In the above the bold text indicates the commands that the user types into the shell.

The commands in this session merely inspect what's available at a given level and then select one of the elements available to illustrate the kind of investigation that can be performed. The last command is an example of printing out a very simple Xpath which matches all the entries at the location in the HTML file.

Although we could, in principle, continue our investigation in this way until we somehow come across what we're looking at there's a more structured way to achieve this.

First we return to Firefox and use the 'Inspect Element' menu item on one of the times we think is in the table we're interested in. (Notice that the path is listed at the bottom.)

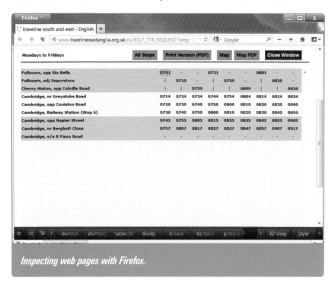

Inspecting web pages with Firefox.

Next track backwards along the path to the first table definition that surrounds this time and right-click on it.

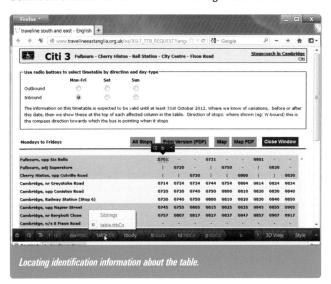

Locating identification information about the table.

Here we see that the table has an associated 'class' of 'ttbCo' and we can use this, rather than the full path to that point, to identify the table in an XPath expression. The expression in question is:

```
//*/table[@class="ttbCo"]
```

which will find all of the tables in an HTML document (at any depth) whose 'id' is 'ttbCo'. (We hope there will be only one.) The '//*/' part of this expression means 'find the following at any depth' after '/' (which is the root). Unfortunately if we try 'cd'ing to this location in xmllint we get:

```
/ > cd //*/table[@class="ttbCo"]
//*/table[@class="ttbCo"] is a 6 Node Set
```

That is, there are actually six tables with that class, so we need to refine our path. (The six tables are the inbound/outbound weekday, Saturday and Sunday versions of the timetable.)

Going back to Firefox we can see that just before the table there's a <div> element which has an 'id' of 'ttbCo_R_1' so we can try using that as part of our XPath as well:

```
//*/div[@id="ttbCo_R_1"]/table[@class="ttbCo"]
```

which works, selecting only one table. We can use the 'save' command in xmllint which saves the current part of the document to a file. Our session would look like this:

```
/ > cd //*/div[@id="ttbCo_R_1"]/table[@
class="ttbCo"]
table >save table.html
```

If we use the 'pwd' command we can see that the path to this spot is actually

```
/html/body/center/div/table/tr/td/table/form/
tr/td/div[2]/div[12]/div[6]/table
```

but since the XPath we found is shorter we can use that instead.

However, the point of this section is to find ways of extracting information without using an interactive session. Luckily 'xmllint' provides a command-line argument '--xpath' which selects only that part of the document and writes it to the output. Our command in Bash, therefore, could become

```
xpath='//*/div[@id="ttbCo_R_1"]/table[@
class="ttbCo"]'
xmllint --html --xpath '//*/div[@
id="ttbCo_H_3"]/table[@class="ttbCo"]'
timetable.html > table.html 2>/dev/null
```

(all on one line). The output of a command can be directed to a file using the 'greater than' character (>). The characters '2>/dev/null' are similar but refer to the error stream that is written to a file called /dev/null, which always exists and just discards what's written to it (meaning that the error messages, which we don't really care about, will be thrown away).

Now we have the table we wanted in a file ('table.html') we can use details from it in our own script. In particular we can pick out the first row of the first column from the saved HTML file using another xpath '/table/tr[1]/td[1]'. In fact this cell includes more HTML surrounding the number we want:

```
pi@raspberrypi ~ $ xmllint table.html --xpath
'/table/tr[1]/td[1]'
<td class="ttbCo">
<p class="ttbCo">
<b>0847</b>
</p>
</td>
```

which suggests that the path '/table/tr[1]/td[1]/p/b' would be better. But if you try this you'll get

```
<b>0847</b>
```

To remove the formatting information (<b> is for bold text) you can apply the 'string' function built in to xmlpath, so that the path becomes

```
string(/table/tr[1]/td[1]/p/b)
```

Bash script

Here, then, is the whole Bash script to implement a new command that will print the time of the first bus from Fulbourn to Cambridge Railway Station:

```
#!/bin/sh
page_url="http://www.travelineeastanglia.org.
uk/ea/XSLT_TTB_REQUEST?language=en&command=
direct&net=ea&line=20003&sup=B&project=y08&
outputFormat=0&itdLPxx_displayHeader=false&
itdLPxx_sessionID=EMEA_98_21772312&lineVer=
1&itdLPxx_spTr=1"

table_time_row=1
table_time_column=1

file_page=buspage.html
file_table=timetable.html
xpath_table='//*/div[@id="ttbCo_R_1"]/table[@
class="ttbCo"]'

get_timetable() {
    if curl --silent --user-agent "Mozilla/
    5.0" --output $file_page "$page_url"; then
        xmllint --html --xpath "$xpath_table"
        $file_page >$file_table 2>/dev/null
        return $?
    else
        return 1
    fi
}
read_table() {
    local row=$1
    local column=$2
    xmllint $file_table --xpath "string(/table/
    tr[$row]/td[$column]/p/b)"
    return $?
}
if get_timetable; then
    if time=$(read_table $table_time_row
    $table_time_column); then
        echo "First bus time is: $time"
        exit 0  # success
    else
        echo "Failed to find entry ($table_
        time_row,$table_time_column) in the
        table">&2
        exit 2
    fi
else
    echo "Failed to read the bus timetable
    web page">&2
    exit 1
fi
```

Testing this out we get:

```
pi@raspberrypi ~/wget $ ./first_bus
First bus time is: 0701
```

Extracting information like this would, for example, enable a time to be calculated for a cron entry (described in its own section) that reboots Raspberry Pi at that time (which may have the indirect effect of turning on an attached television if it has an HDMI connection). But you're sure to be able to find many other uses for information such as this as part of your own Raspberry Pi gadgets.

Using Python

Although the command was executed in a Bash shell the same command can be executed in Python by using 'system' command from the standard 'os' module. For example, the xmllint command to extract the table might be implemented as

```
import os

file_page='buspage.html'
file_table='timetable.html'
xpath_table='//*/div[@id="ttbCo_R_1"]/table
[@class="ttbCo"]'
```

```
cmd_xmllint="xmllint --html --xpath '%s' %s
>%s 2>/dev/null"

rc = os.system(cmd_xmllint%(xpath_table,
file_page, file_table))
```

There is however, a whole library in Python devoted to web scraping called 'BeautifulSoup', which provides similar capabilities of XPath but in terms of Python operations.

Help

You can find more out about xmllint using the command

```
man xmllint
```

The syntax of an XPath is very flexible and allows a very wide range of parts of an HTML or XML document to be isolated and saved. It's documented at

```
http://www.w3.org/TR/xpath
```

If you're interested in using the Python BeautifulSoup module you can find more about it on the Web at

```
http://www.crummy.com/software/BeautifulSoup/
```

SOFTWARE RECIPES
Using the Internet

Communication occurs at almost every scale in technology. Information is transferred at huge speeds between separate blocks inside a chip and at progressively lower speeds between the chips on a computer's printed circuit board, cards plugged into it, devices plugged into them and other computers outside it. Once outside the box communications run fastest to other computers in the home and less fast to those outside. At every level the same communication issues can appear. What happens when data is corrupted? What happens when different elements can't keep up with the speed of the data? How should information be represented so that it's commonly understood?

OSI and the Internet

At the turn of the 1970s the US Department of Defense's Advanced Research Projects Agency started work on an ambitious plan to connect computers in different cities together in a network called ARPANet. It took ten years before it joined no more than about 200 computers, but since these were owned by commercial and academic institutions at the forefront of communications research it generated a substantial body of fundamental research and experience.

ARPANet averted much of the potential for chaos with different operating systems, processors and communications mechanisms by obliging every participating computer to have

its own Interface Message Processor (IMP) that undertook the actual communications, but as the sites wanting to join grew, and the capability of computers in general increased, even this level of equipment standardization became unwelcome.

One consequence, particularly favoured in Europe, was a large amount of work in the International Organization for Standardization (perversely, 'ISO') directed at providing standards for Open Systems Interconnection (OSI). The resulting plethora of standards were all related by one 25-page 'architectural' standard ISO 7498 providing many useful concepts and explaining a new way of thinking about a computer network. It involved imagining a network of 'entities' that have to provide an improved communication service to its users while using a lesser imaginary network that also provided services. The resulting stack of imaginary network 'layers' started with a Physical Layer at the bottom, which provided some means to represent bits electrically, and ended with an Applications Layer, where the communicating entities were programs that did useful things for real users. Altogether seven layers were formulated so the standard is often called 'the seven-layer model'.

Today the most important protocols you find implemented in Raspberry Pi evolved organically from those used in ARPANet (which became the Internet once the IMPs had gone). Nonetheless, it's the seven-layer model that's normally used when describing the relationship between the different layers that exist, even in the Internet Protocols.

The protocols

Ethernet and WiFi

The second layer of the OSI seven-layer model is the Data Link Layer, in which packets of data ('Service Data Units', or SDU) can be sent from one entity to its peers, which will be received uncorrupted or not at all. In Raspberry Pi's world the protocols providing a data link service are those provided to access the Ethernet Port or a WiFi adapter if it's fitted.

To see all of the Data Link Layer networks that your Raspberry Pi is connected to try

```
ifconfig
```

which will list the interfaces that have network connections.

IP

The data link service is used by the third layer, the Network Layer, which among other things has the modest aim of providing a literally global network. A packet of data can be sent that may end up anywhere else on the planet.

The protocol providing this service in Raspberry Pi is the one that defines the Internet, called Internet Protocol, or IP for short.

IP addresses

Although not true of all Network Layer protocols, IP was envisaged to have global addresses – the same address referring to the same computer no matter where it's used in the network. Sadly, if your ISP follows the usual pattern, it provides your house with only one IP address, which has to be shared by all the computers within it. This will mean that the IP addresses you use at home don't have this global feature.

The addresses themselves consist of four byte-sized numbers normally written separated by dots (.), sometimes called a 'dotted-quad'. With a total of 32-bits there are over 4,000 million of these addresses, and when ARPANet designed the protocol the thought that this would be insufficient was inconceivable. Unfortunately these addresses have now all been given away.

For several years the problem of running out of addresses has had a solution, namely a new version of IP. The version of IP with 32-bit addresses is version 4; the new one is called version 6. IPv6 supports 128-bit addresses, but because your home probably uses a private network (essentially an Internet just for your own house) with a fairly modest number of computers you probably don't need this feature.

Routers

The network devices that connect one sub-network to another are called Network Layer relays in OSI and 'routers' by the IP specification. The main function of ARPANet's original IMPs was to act as a router. One IP packet will normally be sent from router to router along a path that eventually ends at the Network Layer entity with the given IP address.

Each router has its own IP address and you can use this command on Raspberry Pi to see which routers are used on the way to the IP address for google.com:

```
traceroute 173.194.34.136
```

You should see a chain of a dozen or so routers, the first being one in your own house.

You can see the address of the first of these routers in the rather cryptic output of this command:

```
route
```

This first router to the rest of the world is called a 'gateway'.

DNS

You may notice while doing this that some of the routers are given names, albeit rather obscure ones. Almost every time you refer to an IP address you'll use a name rather than a dotted-quad. In the case above you could have used 'google.com' instead of '173.194.34.136'.

Perhaps the most common use of an IP name is in the Universal Resource Locator (URL) that's used to name a web page. The names are used only as a convenience. A dotted-quad can almost always be used instead. For example, you'll find a familiar search engine appear if you type this URL into a browser:

```
http://173.194.34.136/
```

There's a truly global network service, the Dynamic Name Service (DNS), whose millions of servers exist on the Internet to provide the IP address that corresponds to any given IP name. To see these addresses on Raspberry Pi try, for example,

```
host facebook.com
```

UDP and TCP

The OSI Transport Layer is the next layer up the stack from the Network Layer. It's all about providing the user with what they want rather than what the network wants to give them.

OSI envisaged a choice of two kinds of service at almost every layer: one connection-orientated, where data can be streamed after a connection has been opened and then the connection is eventually closed; and connectionless, where no open and close is necessary – data is simply sent. IP provides the simpler of these two kinds of service.

If the user wants a connectionless service or a connection-orientated service they can have it based on whatever the Network Layer provides. If the user wants a particular size of SDU the Transport Layer will provide it no matter what size the packets are in the Network Layer, and so on.

The Internet supplies two protocols, one providing a connectionless service and the other a connection-orientated one.

UDP

The simplest of the two is the connectionless User Datagram Protocol which sends just one packet of data of any size. In point of fact the maximum quantity of data is larger than the memory that would have been available in the IMPs and so may originally have been considered to be 'infinite'. Today it looks more modest but is still over 64,000 bytes.

TCP

Unlike UDP, the connection-orientated Transmission Control Protocol (TCP) provides a service that includes data integrity, guaranteeing that – barring catastrophe – data sent will be received. In order to provide this service it may have to transmit some data more than once, which implies that it cannot completely guarantee how long it will take to transport any specific item. This is very often no problem at all, but occasionally, when time-critical information such as live video or audio has to be carried, this feature means that UDP may be preferred.

Ports
Both TCP and IP use the same kind of Transport address, consisting of the IP address together with a small integer called a 'port' number that distinguishes the particular Transport entity within the computer. For example, when your web browser is downloading pages and pictures from many different web addresses at once each separate connection will have its own port number to distinguish it from the others.

When your computer is providing a service to other computers (we discuss quite a few of these in other sections) it'll probably have its own well-known port number that it'll listen to for data containing requests.

Many services have been registered globally with a specific port number to use like this. For example, a web server will normally use port 80.

Use this to see the other port numbers your Raspberry Pi knows about:

```
less /etc/services
```

Sockets
Unix-like operating systems (including Linux) all have the same set of C functions for using IP, TCP and UDP (and almost every other protocol) that was originally created in the University of California at Berkeley. This Application Programming Interface is called the 'Sockets' API, and the same set of functions have been adopted, with a certain degree of success, in the Windows operating system in the form of the 'Winsock' API. Although we describe it in C, the same API is available in many other languages, including Python.

Unsurprisingly the functions seek to provide a plug-and-socket analogy. An established connection is like a socket with a plug inserted into it. Generally this will mean that a connected socket has two addresses associated with it: the local entity address (an IP address and port number in the case of UDP or TCP) and a remote entity address. You might think that including the local IP address as part of a socket definition is a bit redundant, because Raspberry Pi will have only one IP address, but this isn't so: for example, you might have a connection to the Internet through both Ethernet and WiFi and both of these interfaces will have their own IP addresses.

Getting address information
Because the same socket functions are used for many different protocols (IPv6 has its own version of TCP and UDP, for example), ways are needed to specify different 'address families' (for example, the C constant 'AF_INET' refers to IPv4 addresses, whereas 'AF_INET6' refers to the much larger IPv6 addresses), and different 'protocol families' (for example the IPv4 version of IP, TCP and UDP are referred to by 'PF_INET', and IPv6 by 'PF_INET6').

A socket specifies the type of service its protocol must

provide, not the actual protocol itself. 'SOCK_STREAM' specifies a connection-orientated service, which will result in TCP being used; and 'SOCK_DGRAM' specifies a connectionless service, which will result in UDP being used. Because the API makes it so similar to use either protocol many Internet services are implemented in a way that allows you to use either.

The first C function that you might use is 'getaddrinfo' which allow you to use textual names to establish an address. It uses one string as an IP name (which might be looked up in DNS) and another string to name a port (which might be looked up in /etc/services). Because there might be more than one protocol available (ie there might be both UDP and TCP connections) the function actually returns a list of possible addresses, although you can provide a specific address requirement.

In this example we're looking for a TCP address for the 'http' port at the computer with IP name 'www.google.com':

```
#include <sys/types.h>
#include <sys/socket.h>
#include <netdb.h>

struct addrinfo required;
struct addrinfo *list;
int rc; /* a return code we hope will be zero */

memset(&hints, 0, sizeof(required));
required.ai_family = AF_UNSPEC; /* any address
 family */
required.ai_socktype = SOCK_STREAM; /*
connection-orientated */

rc = getaddrinfo("www.google.com", "http",
&required, &list);
```

If we wanted to insist on an IPv4 address we would replace 'AF_UNSPEC' with 'AF_INET'.

Because the list of addresses is placed in its own specially allocated region of memory it needs to be freed when it's no longer needed. The C function to do that is 'freeaddrinfo':

```
freeaddrinfo(list);
```

Creating a socket
It's possible to work through the list of addresses provided, but many people simply accept the first. The addressing information held in a 'struct addrinfo' includes an address family (eg 'AF_INET'), a socket type (eg 'SOCK_STREAM') and a protocol. The protocol will be represented by one of the numbers contained in the file /etc/protocols. This is enough information to create a partly filled-in socket using the 'socket' function:

```
struct addrinfo *address;
int sockfd;
address = list; /* Take the first in the list */
sockfd = socket(address->ai_family,
                address->ai_socktype,
                address->ai_protocol);
```

'sockfd' is the number of a file descriptor, the same kind of thing that's used to read and write data to a file.

Establishing a connection

There are two main ways for a socket to become connected: either you'll provide a socket with no plug in it or you'll provide a plug that you'll insert into someone else's socket. Typically 'servers' do the former and 'clients' to the latter.

Plugging in as a client

To create a connection as a client is a simple matter of taking the socket file descriptor and the address we want to use and calling the 'connect' function:

```
int rc; /* a return code we hope will be zero */

rc = connect(sockfd, address->ai_addr,
address->ai_addrlen);
```

Naturally if there's no socket to be plugged in the connection attempt will fail and this will be indicated by a non-zero return code from 'connect'.

Providing a socket as a server

Acting as a server can be a little more complicated than acting as a client.

To act as a server you must first claim a local address to use for your new socket. Normally you'll want to reserve Transport addresses on all the networks the computer is attached to, but you may occasionally want to use only the address corresponding to a specific interface. The information about which addresses to 'bind' to in this way is again provided by a 'struct addrinfo', and you could, again, fill in its details using 'getaddrinfo':

```
struct addrinfo required;
struct addrinfo *list;
struct addrinfo *myaddress;
int rc; /* a return code we hope will be zero */

memset(&hints, 0, sizeof hints);
required.ai_family = AF_UNSPEC; /* any address
family */
required.ai_socktype = SOCK_STREAM; /*
connection-orientated */
required.ai_flags = AI_PASSIVE; /* fill in my
own address details */

rc = getaddrinfo(/*my address*/NULL, "4321",
&required, &list);

myaddress = list; /* Take the first in the list
*/
```

This example would provide an address that can be used for the 'here' side of a socket, which will require a connection-orientated (TCP) connection to local port 4321.

To reserve this address for your application you can now use

```
int rc; /* a return code we hope will be zero */

rc = bind(sockfd, myaddress->ai_addr,
          myaddress->ai_addrlen);
```

The code said it didn't care about which address family was used, but this call will work equally well whether 'myaddress' turns out to be an IPv4 address or an IPv6 one. Naturally if some other program has already claimed this address the bind will fail and you may have to choose another port number. You therefore have to take some care about the port number you request. Port numbers below 0xC000 (49152) can be reserved for your program by the Internet Assigned Numbers Authority (IANA). Numbers larger than this can be used, on a temporary basis, by anyone. Port numbers less than 1024 can only be used by programs that are run by the privileged 'root' user.

Once you have exclusive use of the port you can wait for someone to initiate a connection to it using 'listen':

```
int rc; /* a return code we hope will be zero */

rc = listen(sockfd, /*backlog*/1);
```

This will wait, perhaps forever, until a connection comes in. Like London buses you may find that you wait for ages before one arrives and then several all turn up at once. The backlog (1 in the example) is the maximum number of connection requests that the system will hold for you while you're handling this one (each will eventually need its own 'listen' to be received). When the number of pending connections is greater than this number they're rejected automatically.

It may be some time before your program has finished with this connection, which could be irritating to potential users of the service. To solve this problem there's one further call in the sockets API which allocates a dynamic port and moves the incoming connection on to it. This will free the original file descriptor for dealing with the next connection with 'listen' again. This call is 'accept':

```
socklen_t theiraddress_size;
struct sockaddr_storage theiraddress;

theiraddress_size = sizeof(theiraddress);
serverfd = accept(sockfd, &theiraddress,
&theiraddress_size);
```

If an error occurs 'serverfd' will have a negative value.

As well as moving to a new port and creating a new file descriptor (which must be used from now on to talk to the client) this call can also return the address of the 'other end' and the size of the address structure (which will probably be appropriate for either IPv4 addresses or for IPv6 addresses).

Sending and receiving data

The way data is sent and received is different depending on whether the file descriptor refers to a connectionless or a connection-orientated service.

Connection-orientated transfers

In a connection-orientated service (such as one provided by a TCP connection when SOCK_STREAM was used) there's a context surrounding the transfer that indicates which addresses are being used, so that information isn't needed in each send or receive. In both cases the data to use and some 'flags' are required along with the socket file descriptor. This is how sending works:

```
char *data = "Raspberry Pi over TCP";
size_t bytes_to_send = strlen(data);
ssize_t bytes_sent;
bytes_sent = send(sockfd, data, bytes_to_send,
/*flags*/0);
```

If an error occurs 'bytes_sent' will have an otherwise impossible negative value. The 'send' may need to be repeated on the later part of the data because the number of bytes sent is allowed to be less than the number requested.

Receiving works in a very similar way:

```
char data[2000];
ssize_t bytes_received;
bytes_received = recv(sockfd, &data[0],
sizeof(data), /*flags*/0);
```

If an error occurs, 'bytes_received' will again have a negative value. If the connection was closed it may have the value zero.

Connectionless transfers

In a connectionless service (such as the one provided by UDP when 'SOCK_DGRAM' was used) there's no context that determines the addresses that should be used in each transfer, so addressing information has to be included in the send and receive functions used. The two functions to use are 'sendto' and 'recvfrom', which are closely based on 'send' and 'recv'.

This is how sending works when transmitting data to a particular IPv4 destination:

```
char *data = "Raspberry Pi over TCP";
size_t bytes_to_send = strlen(data);
ssize_t bytes_sent;
struct sockaddr_in destination;

/* set up destination */

bytes_sent = sendto(sockfd, data, bytes_to_
send, /*flags*/0,
        &destination, sizeof(destination));
```

The value assigned to 'bytes_sent' is the same as the analogous call to 'send'. To send to an IPv6 address 'destination' would have to be a 'struct sockaddr_in6' instead of a 'struct sockaddr_in'.

Receiving works in a very similar way:

```
char data[2000];
ssize_t bytes_received;
socklen_t source_size;
struct sockaddr_storage source;
int rc; /* a return code we hope will be zero */
source_size = sizeof(source);
bytes_received = recvfrom(sockfd, &data[0],
sizeof(data), /*flags*/0);
```

If an error occurs 'bytes_received' will be set as it was in 'recv'. The address of the peer entity that sent the data will be written

in to 'source', which could be used as the destination in a subsequent 'sendto' in order to send a reply.

Closing the connection

Once your program has no further use for a connection it can be closed using the 'close' function, which works on any file descriptor, not just those used for sockets.

```
close(sockfd);
```

Help

All of the calls we've talked about are described in even more detail on your Raspberry Pi in 'man' pages. This includes getaddrinfo, freeaddrinfo, socket, connect, bind, listen, accept, send, recv, sendto, recvfrom and close.

If you want a huge amount of additional information one of the best sources is the book *UNIX Network Programming Volume 1, Third Edition: The Sockets Networking API* (ISBN 0-13-141155-1) by W. Richard Stephens with Bill Fenner and Andrew M. Rudoff.

Using sockets in Python

Python has its own module 'socket' which provides more or less identical functions to those described in C above.

In addition to several others it includes functions getaddrinfo, socket, connect, bind, listen, accept, send, recv, sendto, recvfrom and close. It doesn't implement freeaddrinfo because it isn't required in Python.

In addition to plenty of information being available online the following can be used to print information directly from Python:

```
import socket
help(socket)
```

Using sockets in Bash

There are one or two 'pseudo-devices' in Bash that look as if they live in the /dev directory but which don't really exist in Linux. A well-known one is /dev/null, which is empty when read and discards anything that's written to it. Two lesser-known devices are

```
/dev/tcp/<IPaddress>/<port>
```

and

```
/dev/udp/<IPaddress>/<port>
```

which can be used to read bytes coming from a remote connection or write bytes to one.

For example, on a remote Linux computer the 'net cat' command 'nc' can be used to listen on a TCP port (4321, say) and to write anything it receives on that port to a file called 'pi-dmesg', like this:

```
nc -l -p 4321 > pi-dmesg
```

If this machine was, say, at address 192.168.104.66, Raspberry Pi could send the contents of its kernel log (which is printed using the 'dmesg' command) using

```
dmesg > /dev/tcp/192.168.104.66/4321
```

The same devices can be used to send and receive messages to well-known service ports, such as 80, for HTTP.

SOFTWARE RECIPES
Access via a Serial Line

Your Raspberry Pi can send and receive text to and from other computers using a device that has a very long history. Called a Universal Asynchronous Receiver/Transmitter (UART), this only really needs three wires to work – and it needs virtually no software to drive it. This means that all kinds of embedded computer systems provide one, even if it's only intended to be used by the software developers who bring the platform to life.

Often computer systems provide UARTs that aren't actually connected to anything once the system is in full production, and Raspberry Pi isn't really an exception.

Raspberry Pi actually has two UARTs available, one the standard AMBA device shared with VideoCore and the other (normally unused) an ARM-only UART. The ARM-only one isn't normally included in the kernel configuration. We include some information about enabling it at the end of this section.

Freeing the UART

Later on we describe how to wire up the UART on Raspberry Pi so it can be used from a different computer. Even if you've only just added the hardware that would allow the UART to be accessed you'll find that Linux already has plenty of uses for it:

- As a kernel console (to which kernel start-up messages are sent).
- As a user terminal (providing a logon prompt and then a Bash shell).

If you have a hardware project that uses the UART you might want to free it from both of these uses.

Kernel console

The devices that are used by the kernel are nominated by the kernel command line, which is usually set in the file 'cmdline.txt' on the boot partition (which you can usually see from Linux as '/boot'). This file, whose content is all on the same one line, contains a series of items that configure the kernel's boot process. This includes the specification of one
or more devices that will be used as consoles.

This is a normal example:

```
pi@raspberrypi ~ $ cat /boot/cmdline.txt
dwc_otg.lpm_enable=0 console=ttyAMA0,115200
kgdboc=ttyAMA0,115200 console=tty1 root=/
dev/mmcblk0p2 rootfstype=ext4
elevator=deadline rootwait
```

There are two consoles that are set up in this example:

```
console=ttyAMA0,115200
```

and

```
console=tty1
```

The first uses the UART ('ttyAMA0' is its device name) and sets its speed to 115,200 'baud'. The second uses the virtual terminal allocated on the main display (which is the one you'll normally see). To prevent this use of the UART edit this file:

```
sudo editor /boot/cmdline.txt
```

(We discuss how to select your favourite editor in the section about running programs regularly.) Remove 'console=ttyAMA0,115200'.

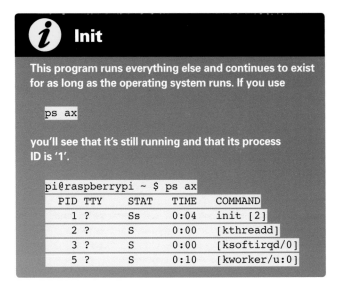

Init

This program runs everything else and continues to exist for as long as the operating system runs. If you use

```
ps ax
```

you'll see that it's still running and that its process ID is '1'.

```
pi@raspberrypi ~ $ ps ax
  PID TTY      STAT   TIME   COMMAND
    1 ?        Ss     0:04   init [2]
    2 ?        S      0:00   [kthreadd]
    3 ?        S      0:00   [ksoftirqd/0]
    5 ?        S      0:10   [kworker/u:0]
```

UART login

When the kernel has finished finding all the devices it can it runs its first program '/sbin/init' from the root file system.
At the bottom of this file you should find the lines

```
#Spawn a getty on Raspberry Pi serial line
T0:23:respawn:/sbin/getty -L ttyAMA0 115200
vt100
```

which will read and write to the serial port /dev/ttyAAM0 (our UART) and set its speed to 115200 baud.

To remove this use of the UART remove or comment out this line. (To comment it out insert a hash (#) at the front of the line starting 'T0:'.)

```
sudo editor /etc/inittab
```

UART speed

The annexe about configuration gives a few options that can be used to change the speed of the UART when it's first booted.

If your project needs the UART to remain at the same speed over a reboot you'll have to take into account that its speed is potentially set to different values at each of the following three times:

▪ Just before the Linux kernel is read from the SD card, to a value potentially contained in /boot/config.txt.

▪ As the kernel comes up and writes text to its consoles, to a value specified on its command line in /boot/cmdline.txt.

▪ As 'init' begins to run Linux and invites logons over terminal devices, to a value specified in /etc/inittab.

Terminal control in C

A UART is one of several sources of character input and output that Linux can treat as a 'terminal'. Terminals have special characteristics that tie together the characters that are typed into them with the characters that are sent back (and displayed).

For example, when you type a character it's 'reflected', with normally the same character being sent back for display. Some special characters, though, have special functions. The Ctrl-D character (produced by pressing the 'Ctrl' and 'D' keys at the same time) is used to mean something along the lines of 'and that's the last thing I'm going to type to you'. The Ctrl-C character will interrupt the current program and may cause it to halt.

Some of the characters, like backspace and delete, Ctrl-U, Ctrl-A, Ctrl-E, forward arrow, back arrow and so on, let you edit the line you're currently typing and it isn't until you send the 'newline' character (sent by pressing the return key) that the resulting edited line is sent to the program that requested the line. On top of that characters are sent during line editing operations (such as deleting a character in the middle of the line) which effectively redraw the whole of the edited line.

Programs that need a line at a time will find this handling of terminal characters very convenient. Programs that need every character as it's typed, such as text editors, wouldn't be pleased if they received nothing until return was typed. The characters sent to echo the line being edited might also not welcome it. For this reason terminals can be set into either 'line mode' or 'raw mode' to suit your requirements.

If your project needs text to be uninterpreted it may need to set the terminal into raw mode.

In addition to these issues you may want to change: the speed at which the UART operates; whether to help guard against corruption by providing 'parity bits' on each character sent or transmitted; or, how a sender of characters can be told that the receiver's going too slowly to keep up (if there's no control of the flow of characters some are likely to be lost and never read).

All of these aspects of terminal operation, including assigning particular functions to different characters, can be controlled by a dozen C functions grouped together under the heading 'termios', which is described in great detail in the 'man' command:

```
man termios
```

To use these functions in C you'll need to start your program with the lines

```
#include <termios.h>
#include <unistd.h>
```

Most of these functions require a 'file descriptor' for an open terminal device, as their first argument. On Raspberry Pi the name of the UART terminal device is

```
/dev/ttyAMA0
```

To provide this file descriptor this device must be opened. If you don't want to respond to requests to interrupt the running program (such as caused by Ctrl-C) coming from the terminal while your program executes, include the 'l O_NOCTTY' part of the program below.

```
int fd; /* file descriptor — a small integer */

fd = open("/dev/ttyAMA0", O_RDWR | O_NOCTTY);

if (fd >= 0) {

    /* Program using termios */

    close(fd);
}
```

These functions are fairly portable across most Unix-like operating systems. In Linux, however, they're implemented by a set of 'ioctl' kernel functions, in the same way that other hardware interfaces are (such as SPI and I²C, which we cover in separate sections). You shouldn't need to use these directly, but you might like to understand that the UART basically works in the same way as the other interfaces. For more information about these use

```
man tty_ioctl
```

UART use from Bash

Simple use

If the UART isn't being used as a console device it should be possible to transmit characters down the serial line provided by the UART simply by using the terminal device as an output. For example, using

```
echo "Hello World" > /dev/ttyAMA0
```

Conversely it's possible to observe characters being sent to the UART by using the device as an input.
For example,

```
cat /dev/ttyAMA0
```

UART control
There's a command that provides access to almost all the functions of the terminal control function 'termios' called 'stty'. Because 'termios' provides access to quite a wide range of features the stty program is similarly complex. To see what it can do use

```
man stty
```

(You might notice a certain similarity to the information obtained using 'man termios'.)
    If not provided explicitly the terminal that stty manipulates is the one attached to its standard input stream. Some people use the slightly perverse idiom

```
stty stuff... < /dev/ttyAMA0
```

(which can only be done by a privileged user) to change the way /dev/ttyAMA0 operates, but it's better to use

```
sudo stty -F /dev/ttyAMA0 stuff...
```

A relatively small amount of experimentation with stty can render your shell useless. To recover from this there's the command

```
stty sane
```

Terminal control in Python
As with many other Unix-style libraries, Python provides an almost literal equivalent of the termios functions in a separate module. To use it you'll have to import the module into your program using

```
import termios
```

Once imported information is available about this module directly from Python itself using

```
help(termios)
```

Raw access by Linux applications
Characters consumed and generated by both devices are 1-byte each, and are the subject of much of the I/O system. The devices can be opened as files in the normal way and can be read when data becomes available (eg using file descriptors and 'select') or written to.
    See the standard Linux 'HowTo' on the topic:

```
http://www.linux.org/docs/ldp/howto/
Serial-HOWTO.html
```

Creating UART devices
Distributions running a device filing system will typically create the /dev/ttyAMA0 device for you (your reference distribution does this). If you build your own distribution from scratch as part of a cut-down embedded system, it may not be created. In that case you may need to create the /dev/ttyAMA0 device yourself.
    When present the (normal) shared UART device has an entry at

```
/sys/devices/dev:f1
```

and when the ARM-only UART is present it can be found at

```
/sys/devices/serial8250
```

It's possible to use the 'mknod' command to create devices from the information contained in these directories. To do this you have to know the 'major' and 'minor' device numbers associated with the device you're creating. These are available (as '<major>:<minor>') in the following file:

```
/sys/devices/dev:f1/tty/ttyAMA0/dev
```

If the above files contained '204:64' you might use this code to create the device '/dev/ttyAMA0':

```
[ -c /dev/ttyAMA0 ] || mknod /dev/ttyAMA0 c
204 64
```

SOFTWARE RECIPES
Access via SSH

You can safely gain access to your Raspberry Pi from another computer by using the Secure SHell (SSH).

How it works

For SSH connection Raspberry Pi must run an SSH server while the computer that you're using must run an SSH client. The connection is normally over the Internet, allowing the two computers to be in completely different locations.

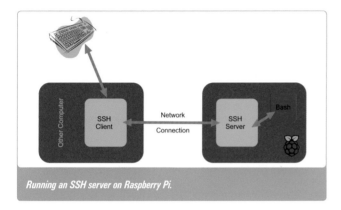

Running an SSH server on Raspberry Pi.

SSH connections are 'secure' because the information that passes between the client and the server is enciphered using a code invented just for that communication. This provides:

- Confidentiality – if other computers on the network receive the SSH data they'll find it impossible to understand; and
- Integrity – if other computers on the network try to change or insert data of their own in place of the SSH data the SSH client and server will know.

Uses

Console connection

The secure shell provides an alternative way to log into Raspberry Pi other than via the console or the UART. The initial part of the procedure authenticates the user to prove that he is who he claims to be, and following that the user is given access to his normal shell (probably Bash) to run a command-line interface.

Remote command execution

Console connection is the normal way SSH is used, but it can be used in other ways too. Rather than stay open for a whole login session it's possible to provide SSH with only a single command line that's run on Raspberry Pi. This can be very convenient in scripts that run on the client computer because it provides a way to run commands on Raspberry Pi just as if they were on the client.

File system access

An SSH connection can also support commands that read and write files in the server's file system. The SSH client normally supports the 'scp' command which can copy files to, from or between SSH servers.

X11 forwarding

SSH also contains a clever mechanism called 'X11 forwarding' that allows an 'X11' service (often simply called 'X') that manages windows on the client to be used by the programs on Raspberry Pi. The effect of this is that if the command run on Raspberry Pi has a Graphical User Interface (eg the web browser or the file manager) the window used will appear on the remote computer. Naturally this is only useful when the SSH client also runs X, so this feature is provided only optionally.

Almost all Unix computers (including Linux computers) provide an X11 service, but Windows computers normally don't. It is, however, possible to install free software on a Windows computer that'll give it the ability to run X and display X11 windows.

Port forwarding

A final use for SSH called 'port forwarding' allows additional connections to be made between the SSH client and the SSH server which pretends to supply a network service at one of them by consuming it at the other. For example, if the client and server were in different people's houses and there were a web server on the house network that the Raspberry Pi was connected to, the SSH client could pretend to be the web server in the remote house (allowing the computers there to use it). Because SSH is used the connection between the two houses will be safe from outside interference and will only be available to people who can prove who they are well enough to set up the SSH connection in the first place. Furthermore it's simpler to protect Internet connections at the two houses because only SSH connections need to be allowed, instead of all the other types of connection (like web server access) that might be needed.

We won't show how port forwarding is set up below but you'll find information about the uses of port forwarding online and can find much of the information you need to use it by typing

```
man ssh
```

Raspberry Pi preparation

Because SSH relies on a network connection you'll have to provide either a wired Ethernet connection or a wireless WiFi connection to your Raspberry Pi first.

Starting the SSH service

You may find that your Raspberry Pi already provides an SSH server. Run the following command to discover whether this is the case:

```
sudo service ssh status
```

If a message like 'sshd is already running' is given, Raspberry Pi is already set up properly and you needn't do any further preparation. If the message is 'sshd is not running' then start the service using

```
sudo service ssh start
```

Also, if you'd like the service to start running every time Raspberry Pi is rebooted you should run

```
sudo insserv ssh
```

If, on the other hand, the message is more like 'unrecognized service' the service will need to be installed from the network. This command should accomplish that:

```
sudo apt-get install ssh-server
```

It should also set the service to run after every boot and start the service immediately.

Finding an IP address
The section about Networking discusses Internet Protocol addresses (four numbers less than 256 separated by dots) and the way they're named in the Internet. Unless you're lucky enough to have a network of computers at home with its own-name server you'll have to name your computer using its IP address, which you can find by typing

```
hostname -I
```

Use from a Unix computer
If your remote computer runs a Unix-like operating system (perhaps it's another Raspberry Pi) you may find that it already contains an SSH client in the form of a command called 'ssh'.

To start a console connection on a Raspberry Pi which has an IP address of (for example) 198.168.0.7 you can use the following command:

```
ssh pi@198.168.0.7
```

This should prompt you for user 'pi's password, which, if you have the reference version of Linux, will be 'raspberry' unless you've changed it (which you probably should).

The conversation ought to look something like:

```
$ ssh pi@192.168.0.7
pi@192.168.0.7's password:
Linux raspberrypi 3.1.9+ #168 PREEMPT Sat Jul
14 18:56:31 BST 2012 armv6l

The programs included with the Debian GNU/
Linux system are free software; the exact
distribution terms for each program are
described in the individual files in /usr/
share/doc/*/copyright.

Debian GNU/Linux comes with ABSOLUTELY NO
WARRANTY, to the extent permitted by
```

```
applicable law.
Type 'startx' to launch a graphical session

Last login: Wed Jul 18 18:01:25 2012 from
192.168.104.120
pi@raspberrypi ~ $
```

– after which you can use Raspberry Pi. You'll be able to return to using the remote computer once you've either typed Control-D (ie pressed the 'Ctrl' and 'D' keys on your keyboard simultaneously) or the 'exit' command.

A single command (eg the one that shows its IP address, 'hostname -I') can be run on Raspberry Pi as follows:

```
ssh pi@198.168.0.7 hostname -I
```

This will look something like:

```
$ ssh pi@192.168.0.7 hostname -I
pi@192.168.0.7's password:
192.168.0.7
$
```

Notice that you need to type 'pi's password every time you run a new command.

X11 forwarding is simply achieved by adding a '-Y' after the command name. From this type of remote computer you're very likely to be running X. To try this feature out you'll need to execute an X-application on Raspberry Pi. The one that displays a file manager is called 'pcmanfm', so try:

```
ssh -Y pi@198.168.0.7 pcmanfm
```

After the normal request for a password you should find that a window appears on your remote machine like this:

pcmanfm running on a remote Raspberry Pi.

You can now use this file manager to explore the file system on Raspberry Pi.

Use from a Windows computer

There are several SSH clients available for Windows:

Cygwin

One is available on the command-line through the Cygwin command 'ssh' and is used exactly as described above. We describe how Cygwin can be installed in the section about connecting to Raspberry Pi through X. Installing Cygwin has the great advantage that it can provide an X11 server for Windows at the same time. To use SSH from Cygwin, run the Cygwin shell and type the command

```
ssh -Y pi@192.168.0.7
```

The '-Y' means that any applications you run on Raspberry Pi that use a GUI (for example Geany) will display on your Windows computer.

PuTTY

Another option, which can be used on its own (if X11 forwarding isn't needed) or together with a Cygwin installation, is to use the free 'PuTTY' application.

PuTTY can be found at this web address:

```
http://www.chiark.greenend.org.uk/~sgtatham/
putty/download.html
```

This is the same program that we describe using in the section about attaching a UART to Raspberry Pi.

Once installed the program can be run by following the menu chain Start > Programs > PuTTY > PuTTY.

This program can connect to different computers using a small variety of mechanisms. It remembers details about each program as a 'session' and before you can use it to access Raspberry Pi using SSH you'll need to create a new session.

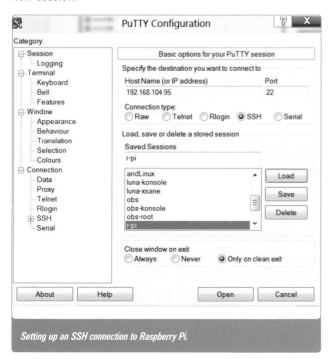

Setting up an SSH connection to Raspberry Pi.

Start by typing a name for your new session (for example, 'R-Pi'), click on the 'SSH' radio button and insert Raspberry Pi's IP address. On the left-hand side of the window there's a panel containing different categories of settings. Click on the plus symbol by the side of 'SSH' and click on 'X11'. Tick the 'Enable SSH X11 forwarding' box if you want to display X programs running on Raspberry Pi.

You should now press the 'Save' button on the PuTTY Configuration dialogue. These settings will now be available whenever you run PuTTY and double-click on 'RaspberryPi' in the list of saved sessions. Which is what you should do next.

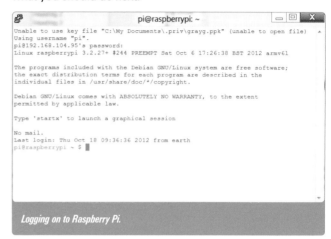

Logging on to Raspberry Pi.

After providing a password for 'pi' you can use the command-line interface. If you know the name of the programs that are called when you click on various parts of the desktop you can call them directly. For example running the command

```
pcmanfm
```

results in a window appearing that looks like this:

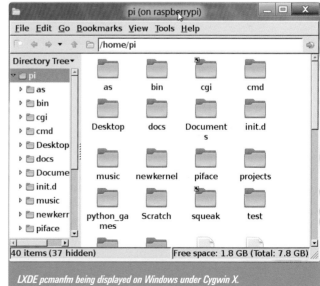

LXDE pcmanfm being displayed on Windows under Cygwin X.

Winscp

Rather than concentrating on console connection the open source program 'winscp' makes use of the access SSH provides to remote file systems. This program can be obtained from

```
http://winscp.net
```

Once downloaded and installed the program can be used to connect to Raspberry Pi. When run you're provided with an interface quite similar to the one that PuTTY provides in which you can set up and store named sessions. The

Making an SSH connection using Winscp.

main details that you should provide are Raspberry Pi's IP address, a user's name and the user's password (eg 'pi' and 'raspberry'). Clicking on 'Login...' will provide a log of the attachment process.

If you accept the default options then once the connection is made Winscp displays a window with two main panels, one showing files on Windows and the other showing files on Raspberry Pi with navigation being possible in both panels. Files can easily be copied from one location to the other by dragging and dropping them either between the panels or between one of the panels and elsewhere on the desktop.

Single commands can also be executed on the remote machine by using Commands > Open Terminal.... and typing into the 'Enter Command' box.

Use outside the home

Although space does not allow the processes involved to be described here in detail it is possible to arrange for Raspberry Pi to be accessed outside your home from an arbitrary Internet location. The steps you will need to perform are:

- Obtain a dynamic DNS entry and a domain name for your home (for example, from http://dyn.com).

- Configure the firewall in the gateway your ISP provided you to allow incoming traffic on the SSH port and direct it to your Raspberry Pi.

Using your new domain name as an IP address you will then be able to your Raspberry Pi anywhere with an internet connection (for example from school, work or a hotel room).

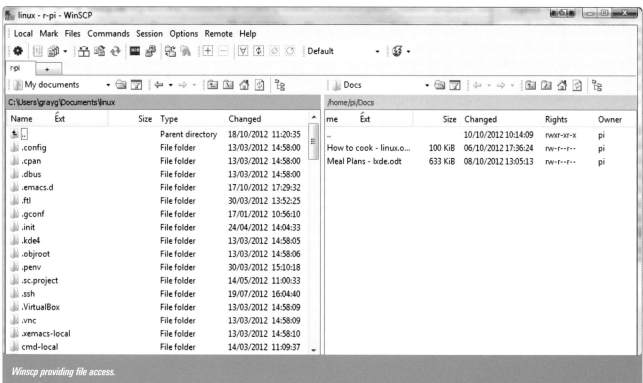

Winscp providing file access.

SOFTWARE RECIPES
Access via X

In the 1980s computers were still very expensive. University departments such as the one at Massachusetts Institute of Technology (MIT) just outside Boston used cheap display terminals to gain access to one or more computers. There was a fair amount of difference between the various computers available so a research program, called Project Andrew, was started to find ways to give students access to this diversity. As part of this project a previous system, called 'W', was used to make the 'X' windows system. X allowed a simple terminal (or 'X-server') to parcel up rectangles of screen area and give them each, along with matching keyboard and mouse events, to different programs ('X clients') running on different computers. The same terminal might display programs (such as text editors or command-line consoles) from many different computers in different windows.

Through the intervening years the history of X has been determined by the liberal nature of its licences and its growth both in terms of innovation (driven partly by its adoption in Linux) and popularity. Although this may finally be coming to an end, X has always been the primary way to use a screen in Unix-style operating systems including Mac OS, but not including Android. Being free software there are versions of X available both on Android and Windows.

The current version of X in use everywhere is version 11 (although the release within that continues to evolve), so the X Windows system is also often called 'X11'.

When you look at a modern desktop computer display, such as one provided by Ubuntu Linux, you'll find a lot of common screen 'furniture' such as scroll bars, title bars, little cross symbols in the corners of windows and special areas of the screen such as the wallpaper and some kind of launch bar. All of these items are built on the windows that X provides by a separate component of the operating system called a 'windows manager'. The look and feel of a desktop is almost completely due to the windows manager in use, not the X windows system that underlies it.

Although X has always been designed to allow the application program behind a window to be supported on a remote computer this feature is used relatively infrequently on today's desktops. Almost all of the X clients are actually supported on the same computer, although there's still no reason why they shouldn't be remote. Conversely there's no reason why any of the windows that appear on a Unix-style desktop shouldn't be displayed on any other computer that runs an X-server.

How it works

In X one computer 'serves' its screen space in individual rectangular windows. Separate programs, such as an editor or a web browser act as 'clients' of the X-server and use it to read and write a graphical user interface to.

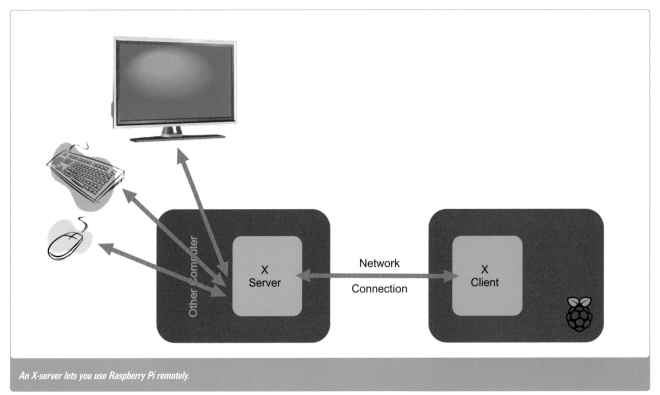

An X-server lets you use Raspberry Pi remotely.

There are two ways that X can be used by Raspberry Pi: one where Raspberry Pi uses X to produce its display on another computer; and one where another computer's display is shown on the television connected to Raspberry Pi. Both of these are useful but this section is going to concentrate on the former, because it's another way to use your Raspberry Pi from somewhere else.

Imagine that you're using your Raspberry Pi to control a weather vane in your loft, or your central heating from the airing cupboard in your house. You won't want to attach a television or monitor to it in either of these cases. Using X you can still use Raspberry Pi's graphical user interface, but from a different computer in your house, or even your mobile phone.

Although X provides a way for an application to use a remote display it doesn't provide any way to run the application in the first place. For that you'll need access to Raspberry Pi's shell. The Secure SHell, SSH, is an ideal candidate and is described separately in another section.

Running an X-server

All the applications that support a graphical user interface on Raspberry Pi already run as X-clients on Raspberry Pi so what we need to do is run an X-server on the remote computer. Currently all Linux desktops run X-servers so there's little to say about using X from these types of operating systems. If the remote system runs Windows, however, you'll need to install an X-server on it.

Installing the Cygwin X-server

In point of fact there are quite a number of X-servers available on Windows, although some of them are commercial products that must be paid for.

The Cygwin project was started by Red Hat (one of the larger Linux distribution maintainers) to provide Unix-style libraries on Windows so that applications written to run on them could be run under Windows. Cygwin provides its own free windows distribution that includes many of the applications familiar to Linux users. Recently this has included an X-server.

Cygwin can be obtained directly from

    http://www.cygwin.com

The X-server has had relevant improvements in recent releases, so if you already had an old version of Cygwin you might wish to update it.

When you install Cygwin you should place the installation program somewhere that you're prepared to return to in the future. You'll need to re-run the program each time you wish to update your installation.

The standard installation doesn't include the X-server or SSH so you must install additional Cygwin packages:

- X-start-menu-icons
- xorg-server
- xinit
- xhost
- openssh

First run

You should find the X-server in your program menu under 'Cygwin-X'. On first execution you may get a message from the Windows Firewall. You should allow the proposed 'exception' (otherwise remote machines won't be able to use the Windows display).

Running the X-server.

Testing your X-server

Launch Start > Programs > Cygwin-X > XWin Server. To connect to the remote machine (which has IP address 192.168.104.95 in this example), from the Cygwin X terminal, type:

    xhost +raspberrypi
    DISPLAY=:0 ssh -Y
        <username>@
        192.168.104.95

The -Y option will tunnel X11 over SSH.

Now, inside the remote machine run an X-using command, for example:

    lxterminal &

should launch a Raspberry Pi terminal window on your local machine.

Allowing hosts to use your X-server

The 'xhost' command above temporarily sets another machine from which the X-server will accept connections. You can make these access controls permanent by placing a list of machines in the /etc/X0.hosts file (for your normal display, :0). For example, my file contains the hosts I need in order to work with Raspberry Pi (from a Cygwin window):

```
% cat /etc/X0.hosts
localhost
192.168.104.95
```

Controlling what runs when the X-server starts

You'll probably notice that a terminal window opens on the top left of your screen when you start the X-server. This is 'xterm', which is run by default. You can start the X applications you wish simply by creating a script '.startxwinrc' in your (Cygwin) home directory that runs them instead of xterm. If you don't want anything to start just create an empty script, eg by typing

```
touch .startxwinrc
```

when xterm opens.

Surviving reboot

It's possible to use the X-applications simply by remembering to run the X-server (if it isn't already running) beforehand. On the other hand you may find it easier to copy the shortcut to the X-server in Start > Cygwin-X > XWin Server to Start > Startup so that it's automatically run on each Windows logon.

Correcting your X DPI settings

When X draws a number of pixels on your screen that are supposed to be a specific size it needs to know how many pixels would measure an inch. It knows this value as the number of dots-per-inch (or DPI). If you find that your fonts are unusually large or small it may be that you need to set the server's DPI. You can do this by changing the program's shortcut for 'XWin server' from

```
C:\cygwin\bin\run.exe /usr/bin/bash.exe -l
-c /usr/bin/startxwin.exe
```

to (for example)

```
C:\cygwin\bin\run.exe /usr/bin/bash.exe -l
-c "/usr/bin/startxwin.exe -- -dpi 90"
```

To make this change right-click on the programs entry for 'XWin Server' and select 'properties'.

One-click access to an X-session

Commands such as a file manager (eg 'pcmanfm') can be run from your quick-start bar.

To do this you need to install PuTTY on Windows. We describe how this can be done in the section about accessing Raspberry Pi using a UART.

Create a new session called, for example, 'build-server' setting:

- Connection type: SSH.
- Close window on exit: always.
- In SSH set the remote command to eg 'pcmanfm'.
- In Connection>SSH>TTY set 'Don't allocate a pseudo-terminal'.
- In Connection>SSH>X11 set 'Enable X11 forwarding'.
- Back at Session type the name of the session, eg 'rpi-files', and save it. Now save a shortcut to PuTTY executable somewhere useful (eg on the quick-start bar) by right-dragging the start-menu entry to the start bar and selecting 'copy'). Rename the shortcut to, eg, 'rpi-files-putty'. Alter the shortcut's properties so that the Target includes a final '-load rpi-files'.

You should find that clicking the new start-bar item will open a window allowing you to authenticate yourself and will then open your chosen X application.

To remove login query in PuTTY

If you haven't already, use PuTTYgen to create a private and a public key. Save the private key somewhere 'safe' where others can't read it. Save the public key file and transfer it to the remote machine to your local file .ssh/id-rsa.pub.

Note: PuTTYgen gives you the option to protect your private key with a password. If you use this feature you'll have to type that password when you use this method (and this is what we're trying to avoid).

On the remote machine (raspberry pi) create or edit the file .ssh/authorized_keys so that it includes a (long) line starting 'ssh-rsa ' followed by the hex contents of .ssh/id-rsa.pub all concatenated on to the same line, optionally immediately followed by a comment starting '=='.

Example line (but all on one line):

```
ssh-rsa AAAAB3NzaC1yc2EAAAABJQAAAIBp65fmesBB
1F868CvE7U9NPQ4bDrN98KC2cJCVrVKAvaM1RrEjL/1jT
MOfs7MH4ssNtQiiYLTxgd7MgfVWXH2P9wggpUk3O+Z4MEH
pOs7aOhD/EfBjyK5wKmIStViizMmI1WWNk7K+/F7yzbsUo
M06uj/zQQtlO2SHUnK945s3nQ==Grays key
```

Check the access controls on the two files in .ssh ('ls -l .ssh') and ensure that only the owner has write access to them: for example,

```
-rw-r-r- 1 pi  219 2010-09-24
17:33 authorized_keys
-rw-r-r- 1 pi  294 2010-09-24
17:23 id-rsa.pub
```

otherwise the public key authentication exchange will be rejected.

Run PuTTY, load the session you want to modify ('rpi-files'?) then, in SSH>Auth, put the private key location into 'Private key file for authentication'. Then save the session.

You should find that the PuTTY terminal appears only briefly before your X application runs. I've found that the PuTTY terminal sometimes stays there, but if you set it to be minimized when run (in the 'shortcut' tab of the shortcut properties set 'Run:' to 'Minimized') it will be less visible.

SOFTWARE RECIPES
Provide a Web Server

When Sir Tim Berners-Lee generalized the existing notion of a 'hypertext' document to try to make information more easily available to the nuclear researchers at CERN, where he was a student, he had international ambitions. Hypertext documents of the time (1990) employed named links that allowed the hypertext browsing program to fetch and display other hypertext documents. His first proposal, which he wrote alongside Robert Cailliau, envisaged using Internet addresses and protocols in hypertext links that literally spanned the world. Until then such links had always been to documents on the same computer. The documents and their links are like the sticky blobs and the strands of silk joining them in a spider's web, so they called the structure they envisaged 'WorldWideWeb'. They couldn't have predicted the extent to which its global references would indeed lead to global relevance. The fact that there's really no need to describe here what purpose there could be in such a web is a powerful testament to the extraordinary success of this concept.

In carrying out the project, Tim Berners-Lee designed the textual language used in a hypertext page, based on the Standard Generalized Markup Language, which was called HyperText Markup Language (HTML), and a new Internet Protocol for copying these documents from one location to another called the HyperText Transfer Protocol (HTTP). He wrote the first HTTP server (now called a web server) and the first 'browser' – an HTML display program that acted as an HTTP client.

In their original proposal they envisaged the construction of an initial web of some three or four computers each costing between £10,000 and £20,000. Today you can build a web browser easily in a £25 Raspberry Pi. The £50,000 they needed for computers could now have bought 2,000 Raspberry Pis.

Why you want a web server

The easiest type of web page to provide on Raspberry Pi is the kind that simply reproduces the same information every time it's viewed. Static pages such as these are as easy to provide as copying a file into Raspberry Pi's web server space, and this can be done in any number of ways from a remote machine, including using SSH and using a Windows Share (both of which we cover in sections elsewhere).

For a gadget enthusiast who's attached all kinds of data acquisition devices to Raspberry Pi (we discuss the use of many hardware interfaces, how information can be accessed from remote locations and 'web scraping' in other sections), pages can be used to view the current set of simple results and offer all kinds of ways to browse more complex results.

Even more excitingly your gadgets and web robots can be directed and controlled through a web server using forms and web page addresses that can use 'CGI' scripts and programs that you write. This section will run through an example of how this can be done.

If you don't fancy writing programs yourself there are several off-the-shelf web applications that you can install on your Raspberry Pi that'll provide useful services, such as hosting your own online web log ('blog'), organising your documents and personal information in a 'content management system', or providing remote access to your collection of music (we describe such an application in a separate section).

For the fastest and easiest way to maintain a set of static pages you can download and run a Wiki application (wiki is Hawaiian for 'fast') where you could do things like provide web pages that describe how to use your computer creations for the rest of the family to read.

How it works

In HTTP one computer acts as a 'client' (the one using a web browser) and the other as a 'server' (a web server). Naturally a web browser can use HTTP connections from many, many web servers, but at any one time one HTML document is loaded only from one web server. A simple server is required to do very little: it holds a repository of hierarchically named files (each document is normally simply a single file in

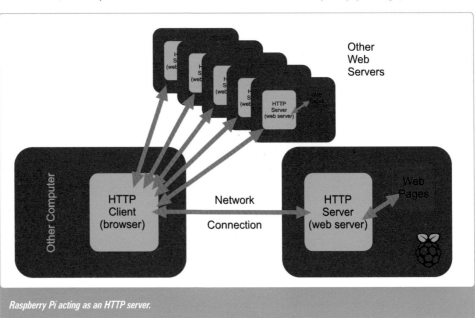

Raspberry Pi acting as an HTTP server.

the server's file system) and it provides its contents when requested. More complex servers write the web page that's returned on each access.

Naturally there are two ways that HTTP can be used by Raspberry Pi: one where Raspberry Pi uses HTTP to browse the Web and, another where Raspberry Pi serves web pages to clients using HTTP. Both of these are useful but this section is going to concentrate on the latter, because it's another way to use your Raspberry Pi from somewhere else. We cover the use of Raspberry Pi as an HTTP client in the section about web scraping.

Not only can you provide information for the web browsers in your own home from a central location on a Raspberry Pi, but you can use the web pages it provides to control your own programs running on it.

Preparation

Although it's quite possible, and very simple, to provide web pages on the same computer as a browser the main point of HTTP is to provide web pages via a network connection, so you'll have to provide either a wired Ethernet connection or a wireless WiFi connection to your Raspberry Pi first.

There are, in fact, a number of web servers available on Linux. One of the most popular, and capable, is called Apache, which is installed simply by typing

```
sudo apt-get install apache2
```

As well as installing the programs to run a new service ('apache2') on Raspberry Pi this will also start them running immediately, and schedule them to start after every reboot too. This command can be used to check whether the service is actually running:

```
pi@raspberrypi ~ $ service apache2 status
Apache2 is running (pid 11180).
```

Apache2 is highly configurable, initially through the text file at

```
/etc/apache2/httpd.conf
```

Consequently that file can be quite complicated, but the version of this file it starts with allows you to start using the server immediately.

HTTP service name – a URL

The section about Networking discusses Internet services and the way they're supported on one of the 'ports' of a computer with given Internet Protocol addresses. One of Tim Berners-Lee's most far-reaching decisions was his design for a Universal Resource Locator (URL), or web address. Although he provided a new protocol to obtain web pages (HTTP) he realized that there were already a number of protocols that would allow a browser to fetch a web page from a computer, so he built the name of the file access scheme, the IP address of the computer and the port it used into a web address. Although the scheme-specific part of a web address can vary, many of them follow the same pattern. The basic idea is to make a URL fit a pattern of characters like this:

- The name of the scheme (eg 'http').
- A colon (:) – what follows is the scheme-specific part.
- Two slashes (//).
- Optionally the name and possibly password of the user logging on to the computer followed by an 'at' sign (@) – if a password's given it follows the name with an intervening colon (:).
- The IP name or address of the computer.
- Optionally a colon (:) followed by the port number.
- A slash (/).
- The name of the web file on the computer starting with a slash.
- Optionally a question mark (?) followed by a query.

To find a URL for Raspberry Pi you'll need to know its IP address, which you can find by typing

```
hostname -I
```

To use the HTTP server the client will use 'http://' followed by this IP name followed by a slash (/) and then the name of a web page. So, if Raspberry Pi's IP address is 192.168.104.95 and there's a web page called 'index.html' (there usually is) the client could use the URL

```
http://192.168.104.95/index.html
```

Your own website

Once you have Apache working you can create your own content to build almost any kind of website. You might like to read the Haynes *Build Your Own Website* manual for ideas. Alternatively there's plenty of information, advice and software available online.

Here, however, we need to get a little closer to the way simple web pages work, to discover how to write programs that will create web pages for us.

HTML files

In Apache, the name of a web page is literally the name of a file in the directory

```
/var/www/
```

although this can be altered later. This means that a web address such as

```
http://192.168.104.95/sittingroom/
using-tv.html
```

will actually use the web page stored in the file

```
/var/www/sittingroom/using-tv.html
```

All web servers will fill in the name of a file if the web page names a directory instead of a file. Apache will usually use the file name 'index.html'. So that the URL

```
http://192.168.104.95/sittingroom/
```

or

```
http://192.168.104.95/sittingroom
```

will use the web page stored in

```
/var/www/sittingroom/index.html
```

and

```
http://192.168.104.95
```

will use

```
/var/www/index.html
```

If we look at this file, for example using the command

```
cat /var/www/index.html
```

we can see that the Apache installation has already put a file there which contains

```
pi@raspberrypi ~ $ cat /var/www/index.html
<html><body><h1>It works!</h1>
<p>This is the default web page for this
server.</p>
<p>The web server software is running but no
content has been added, yet.</p>
</body></html>
```

providing a glimpse at what an HTML file looks like.

Try this out on a web browser (such as Chrome, Firefox, Safari or Internet Explorer) on another computer on the same network as your Raspberry Pi (substituting your own IP address). You should see something like:

The default Apache web page.

This particular web browser (Chrome) allows us to see what the HTML was that it is displaying (settings: tools > view source), which you can see is exactly the same text that was placed in /var/www/index.html:

The HTML source for the default page.

The key to understanding HTML is to spot that it is full of regions of text that begin with a tag like '<something>' and ends with '</something>'. The whole document is between 'html' start- and end- tags, which tell the browser which parts to interpret as the web page definition.

All of the text inside that's delimited by a 'body' start- and end-tag shows which part of the document is supposed to appear in the main part of a browser's display. There could also have been another section using 'title' tags that tells the browser what to call the page.

Inside the body there's one area marked out by the 'h1' tag and two areas marked with the 'p' tag. The 'h1' tag tells the browser to display its text as a heading and the 'p' tells it to display its text as a paragraph. Although the HTML is split up into different lines the browser pays no attention to this layout. The web page won't show text on different lines just because the HTML source is on different lines, for example. In fact the web page would be displayed in exactly the same way even if all of the HTML was written on just one long line.

A first web page

It's easy to make a first web page simply by editing this file. To do that we'll have to pretend to be the privileged 'root' user, because the files under /var/www are read and written by Apache, which runs as root.

The Geany editor was described in the Python section and, assuming it was installed, we can use that to edit HTML files. You'll need to open a command-line interface in the GUI again (using Menu > Accessories > LXTerminal). The command that uses Geany, as root, to edit the index file is as follows:

```
gksu geany /var/www/index.html
```

The 'gksu' command runs the rest of the line as the 'root' user in the same way that 'sudo' does, but does a few extra things that are appropriate when the command uses the graphical user interface. One such thing is to let the user know that it's been called in a way that doesn't need a password. Once read you can remove this message by clicking the 'Close' button.

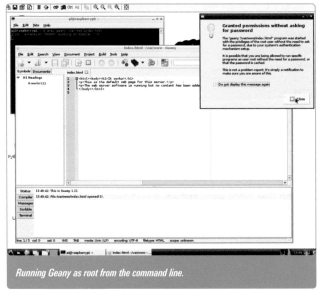

Running Geany as root from the command line.

```
<html>
<title>Mi Pi Page</title>
<body>
    <h1>A Cosmos of Shiney Things</h1>
    <p>Welcome to the universe of useful
    Raspberry Pi web services.</p>
    <p>Here is a complete list of the
      incredible things you can do right from
      this page:</p>
    <ul>
      <li>Read about the stunning things that
        can be done on this page.</li>
    </ul>
      <p>(End of list)</p>
</body>
</html>
```

As you type this you'll notice that Geany will help you out by indenting the mark-up text nicely and providing an end-tag for each start-tag that you type.

You might also notice the introduction of two new pairs of tabs 'ul' (for unordered list) and 'li' (for list item). All the areas delimited by the 'li' tabs are included inside the area marked out by 'ul', and they appear as separate bulleted items. Another tab 'ol' introduces an ordered list, in which list items are displayed numbered.

Save this text (perhaps using the menu entry File > Save) and then test it. You can do this in the same way as before (just reload the page if you still have the browser open). If it's inconvenient to use another computer you can use this URL in Raspberry Pi's normal browser. Then open the URL:

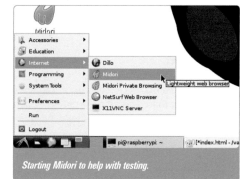

Starting Midori to help with testing.

```
file:///var/www/index.html
```

If everything has worked well you should see:

A first web page.

Naturally you can use the same process to create your web pages elsewhere in /var/www.

HTML tags

It's possible to use a number of word processors to generate HTML directly (usually via a File > Save As... menu), and if you want simply to provide pages of static information this is a good way to get a stock of pages to put on Raspberry Pi. LibreOffice is ideal. You can install it using

```
sudo apt-get install libreoffice
```

This is quite a large program that might require more than the standard 2GB of SD storage. You could also create new pages on other computers and then copy the files into position using one of the many ways to copy files to Raspberry Pi described elsewhere in this book.

A good (and free) program to edit HTML that runs on both Linux and Windows machines is 'kompozer' which will allow text, forms, scripts, and styles to be created in one program.

To write HTML on Raspberry Pi, and to learn the details of HTML properly, it's best to use one of its editors (such as Geany) to write the mark-up language directly.

To do this you might like to know about one or two more tags beyond those introduced above. There's no need to use all of the available tags but if you're interested you can find a definitive list of the tags used in the latest version of HTML at

```
http://www.w3schools.com/tags/
```

This site is an excellent resource because it allows you to experiment with the tags to discover what they can do.

HTML tags are used as a way to

▪ Provide information about the page itself.
▪ Produce headings.
▪ Format text.
▪ Produce tables.
▪ Make links to other web pages.
▪ Display images.
▪ Describe forms.
▪ Include programs and scripts.
▪ Define the style and look of the page.

The importance of tables

A table is normally a sequence of rows each containing the same number of cells so that the cells line up underneath each other to form columns. The text in a cell can be any HTML, including other tables.

Tables are very important in HTML because not only can they be used to provide the usual kind of tables seen in documents, with lines around each cell, but can also provide tables in which the cell boundaries aren't displayed to lay out the text on the page.

For example, the common page layout where there's a column of text on the left with handy information or links to other web pages can be accomplished by making the whole page a table with a row of only two 'cells' in it – one on the left column and one on the right:

```
A line or two at the top of the page.
<table border="0">
  <tr>
    <td width=20%>
      Left hand column of the page taking up
      20% of the page width.
    </td>
    <td>
      Almost all of the web page.
    </td>
  </tr>
</table>
A line or two at the bottom of the page.
```

Web links

Perhaps the most familiar aspect of web pages is the ability to click on something to go to a web page. The anchor tag that provides that facility also provides a different feature that allows an 'anchor' to be defined on the page.

Although most URLs simply refer to a web page it's possible (eg in a very long web page) to go to a specific position within that page. This is achieved by placing name anchors in the page which are then referred to by adding a hash (#) and the anchor name to the end of a URL. For example, if the URL of the page containing this fragment

```
<a id="nicebit">Lovely place to come back to
later.</a>
```

was 'http://192.169.104.95/mywebpage.html' then this URL would go directly to that position on that page:

```
http://192.168.104.95/mywebpage.html#nicebit
```

Forms

Forms are used to gather together information usually provided by a web-page user which is then sent to a web server for processing in some way. They can appear anywhere on a web page, and the same web page might have more than one of them.

There are two ways that the information gathered can be submitted to a web server:

■ By using a special 'post' transaction supported by HTTP.
■ By getting a new web page whose URL has the information from the form incorporated.

These are the 'post' and 'get' methods respectively. The former is the better in terms of reliability and its ability to deal with large quantities of data, but the latter is often the easiest to support. If a 'form' tag's 'method' attribute isn't given the 'get' method is used by default.

When describing a URL above we said that its last section could be a question mark (?) followed by a query. This query consists of a number of 'name=value' pairs separated by ampersands (&). The idea is to support a web page which provides information tailored to the values of the given names. For example, a website that knows a list of numbered jokes and how funny they are might display information about a joke in response to a URL such as

```
http://killerjoke.com/?number=26
```

which might provide information about joke number 26.

If the site needs more information, for example about your likely sense of humour, it might need an additional argument such as the user's country of origin, for example:

```
http://killerjoke.com/?number=35&homeland=france
```

This is an example of a form that might be used to collect this information and request the correct web page:

```
<form action="http://killerjoke.com/"
method="GET">
<fieldset>
<legend>Joke Query</legend>
Joke number: <input type="text" name="number"
/><br />
Homeland: <input type="text" name="homeland"
/>
</fieldset>
</form>
```

CGI scripts

The Common Gateway Interface is a simple standard that allows programs on a web server to be called in response to query URLs, such as those constructed by forms in HTML. Such programs are simply called 'CGI scripts'.

On Raspberry Pi we'd like to use programs written in Python, but the method described below can be used to run programs in any language.

Preparation

Before we can use CGI scripts in Python we need to prepare an area of the file system to hold them and tell Apache about them. These commands will create a place to put scripts

```
/user/share/pi-www/cgi
```

and ensure that both Apache, which can read and write files as the imaginary user 'www-data', and the programmer can share the files placed there:

```
sudo mkdir -p /usr/share/pi-www/cgi
sudo chown www-data:www-data /usr/share/
pi-www/cgi
sudo chmod -R a+w /usr/share/pi-www/cgi/
```

Apache can support more than one website (for example, pretending to have different IP addresses or available on different ports) and keeps separate details for each site in the directory

```
/etc/apache2/sites-available/
```

By default only the 'normal' site at the Raspberry Pi's IP address and active on port 80 is supported, and its details are held in the 'default' file there.

To tell Apache that CGI files are allowed to be executed at our new location we must add this line into that file:

```
ScriptAlias /pi-cgi /usr/share/pi-www/cgi
```

between the initial '<VirtualHost>' line and the ending '</VirtualHost>' line. This can be achieved by editing the file:

```
sudo editor /etc/apache2/sites-available/default
```

Once that's been done we have to ask apache2 to reload all its configuration files so as to read the changes made:

```
sudo service apache2 reload
```

Test the setup

To test our change place a sample script in

```
/usr/share/pi-www/cgi/test.py
```

containing a Python script which prints a simple HTML file to its output. This is an example:

```
#!/usr/bin/env python

import cgi
import cgitb

cgitb.enable()  # prints python exceptions to
the HTML output

print("Content-type: text/html")  # standard
HTML heading

print("""
<html>
  <head>
      <title>Raspberry Pi does CGI</title>
  </head>
  <body>
     <h1>CGI Welcome</h1>
     <p>Hello World</p>
  </body>
</html>
""")
```

The Python library provides a number of functions that'll make it easier to respond to CGI queries (although this script doesn't use them), and the 'cgitb' module provides the 'cgitb. enable' function which will print useful diagnostics in HTML if something goes wrong.

The whole program is really two 'print' statements. The first prints a line that web browsers expect to introduce HTML content. The next prints the whole of the HTML content. Notice the use of the Python construct for a multi-line string which is introduced by a triple quote (""") and then takes each subsequent line as part of the same string until a final triple quote at the start of a line.

Run the program on its own to check that the program is writing the HTML we expect:

```
pi@raspberrypi ~ $ /usr/share/pi-www/cgi/test.py
Content-type: text/html

<html>
  <head>
      <title>Raspberry Pi does CGI</title>
  </head>
  <body>
     <h1>CGI Welcome</h1>
     <p>Hello World</p>
  </body>
</html>
pi@raspberrypi ~ $
```

The most important test, though, is to see whether the program can be run remotely by Apache through the CGI. Substituting your own IP address, try this URL in a browser:

```
http://192.168.104.95/pi-cgi/test.py
```

After a short wait for the program to be run this is the result to expect:

Web page from a first CGI Script.

Responding to a query

This next example creates a new CGI script in

```
http://192.168.104.95/pi-cgi/test-cgi.py
```

that simply prints out the values of two values in a query:

```
#!/usr/bin/env python

import cgi
import cgitb

def main():
    cgitb.enable()  # prints python exceptions
    to the HTML output

    print("Content-type: text/html")  #
    standard HTML heading
    print("""
<html>
  <head><title>Raspberry Pi does CGI</title>
  </head>
  <body>
""")
```

```
       query = cgi.FieldStorage()
       jokeno = query.getvalue("number", "(No
         joke number provided)")
       homeland = query.getvalue("homeland",
         "(Unknown homeland)")
       print("""
<h1>CGI Testjoke</h1>
<p>Request details:</p>
<ul>
  <li>joke: %s</li>
  <li>homeland: %s</li>
</ul>
</body>
</html>
""" % (cgi.escape(jokeno),cgi.escape(homeland)))
if __name__ == '__main__':
     main()
```

Two features of this code to note are the use of Python's string formatting operator '%' and the function 'cgi.escape'.

Python's % operator goes between a string and a list in the same way that the plus operator '+' goes between two numbers. Instead of the sum of the left and right sides the % operator takes the string on the left and replaces each occurrence of '%s' in it with the next item in the list given on the right-hand side. This means that the first '%s' (after 'joke:') will be replaced by the value of 'cgi.escape(jokeno)' and the second will be replaced by the value of 'cgi. escape(homeland)'. The % operator is, in fact, much more capable than this and has a number of alternatives for '%s' that will print out numbers and other objects in different ways.

Several characters have special meanings in HTML, especially less-than (<) and ampersand (&). Imagine that the less-than character was given as part of the value of 'homeland'. The HTML generated would probably have what would look to the browser like a bad tag in it (or even worse a good tag that makes a nonsense of the surrounding HTML). The way HTML gets around this is to replace (or 'escape') these characters with, for example '<' for less-than and '&' for ampersand. For this reason any string that's incorporated into an HTML document needs to be escaped beforehand. The function 'cgi.escape' returns the escaped version of its argument string.

You can test it out with different values for 'number' and 'homeland', for example the URL

```
http://192.168.104.95/pi-cgi/test-cgi.
py?number=35&homeland=Germany
```

should give

An insulting web application

Once you can run programs on Raspberry Pi through a web browser the way is open for implementing all kinds of projects that can be controlled through your own web pages.

If you have a Raspberry Pi doing something useful in the garage contactable only through WiFi then using a web interface may be a very user-friendly way to monitor its state and request it to perform useful tasks.

In the section about Python we described a little program that generated a random insult. By way of an example web application we'll describe how that might be used to provide a vote-for-the-best-insult website.

How the program works

This is a CGI program like the test program above. It uses the 'rudeness' function that comes from our insult program 'docker' which it imports as a module. (The easiest way to ensure this works is to have this program and 'docker.py' in the same directory.)

The 'rudeness' program uses the random number generator to make choices, but the numbers the random number generates aren't truly random – they come from a fixed sequence of 'pseudo-random' numbers. If you returned to the same point in the sequence and asked again for some random numbers you'd get exactly the same ones that you did originally. The random number generator makes this point in the sequence available as its 'seed', which is actually just a big number.

The consequence of this is that if you remember the seed as it is before asking for an insult you'll be able to force 'rudeness' to produce the same insult later by returning the seed to that value. This means that instead of remembering the text of an insult our program can just remember the seed number that was used to generate it – which is easier to use in the program.

The program keeps track of the insults it knows about using a dictionary, 'votes', which is indexed by this seed number and holds the number of votes that insult has had and the text of the insult too. It uses a very useful Python module 'cPickle' which allows any Python value (such as 'votes') to be written to a file and later read back into the program.

By writing 'votes' to a file whenever it's changed and reading it from that file when the program starts the voting information will be maintained while Raspberry Pi is turned off or rebooted.

The program

Each function has information about what it does in the line after its definition, which are highlighted yellow below below:

```
#!/usr/bin/env python

import cgi
import cgitb

cgitb.enable()  # prints python exceptions to
the HTML output

import cPickle
import random
```

```python
# get our insulting program 'docker' and the
  routine 'rudeness'
# that generates a new insult
from docker import rudeness

this_page="voter-cgi.py"
vote_file = "votes.pkl"
votes_none = {}
votes = votes_none # dictionary of
(votes,insult) indexed by id
random_state = random.getstate()

def load_votes():
    "Read in the list of insult votes we have
    stored in the file"
    global votes
    try:
        fileobj = open(vote_file, 'rb')
        votes = cPickle.load(fileobj)
        fileobj.close()
    except IOError:
        votes = votes_none


def save_votes():
    "Write our list of insult votes to the file"
    fileobj = open(vote_file, 'wb')
    if None != fileobj:
        cPickle.dump(votes,fileobj)
        fileobj.close()

def show_top():
    "Print HTML that starts our web page"
    print("Content-type: text/html")  #
    standard HTML heading
    print("""
<html>
  <head><title>Raspberry Pi Insult Voter</
  title></head>
  <body>
""")

def show_bottom():
    "Print HTML that ends our web page"
    print("""
  </body>
</html>
""")

def vote_form_button(insultid):
    """
    Return HTML in a string that displays a
    button for the given vote number
    """
    global this_form
    return (
    '<form action="%s" style="display:inline;">
\n'%(this_page) +
      '<input type="hidden" name="op"
      value="vote"/>\n' +
```

```python
      '<input type="hidden" name="number"
      value="%d"/>\n'%(insultid) +
      '<input type="submit" value="Vote for
      this insult"/>\n' +
    '</form>\n')


def delete_form_button(insultid):
    """
    Return HTML in a string that displays
    a button to delete the given vote
    number
    """
    global this_form
    return (
    '<form action="%s" style="display:
    inline;">\n'%(this_page) +
     '<input type="hidden" name="op"
     value="delete"/>\n' +
     '<input type="hidden" name="number"
     value="%d"/>\n'%(insultid) +
     '<input type="submit" value="Forget this
     insult"/>\n' +
    '</form>\n')


def new_form_button():
    """
    Return HTML in a string that displays a
    button that requests a new insult to
    vote on
    """
    global this_form
    return (
    '<form action="%s" style="display:inline;"
    >\n'%(this_page) +
     '<input type="hidden" name="op" value=
     "new"/>\n' +
     '<input type="submit" value="Get Random
     Insult"/\n>' +
    '</form>\n')


def getspecific_form_button():
    """
    Return HTML in a string that displays a
    button that requests a web page
    showing the insult identified by a number
    supplied in a text box
    """
    global this_form
    return (
    '<form action="%s" style="display:inline;"
    >\n'%(this_page) +
    '<input type="hidden" name="op" value=
    "new"/>\n' +
    '<input type="submit" value="Get This
    Insult number:"/\n>' +
    '<input type="text" name="number"/>\n' +
    '</form>\n')


def new_actions(insultid):
    return vote_form_button(insultid)
```

```python
def stock_actions(insultid):
    return vote_form_button(insultid)+delete_
    form_button(insultid)

def show_votes():
    "Print HTML that displays a table of our
     votes"
    print('<h1>Insults of Worth</h1>')
    print('<blockquote>')
    print('<table border="1"><tr><th>ID</
      th><th>Votes</th>'+
        '<th>Insult</th><th>Action</th></tr>')
    order = votes.keys()
    order.sort(key=lambda vote: votes[vote]
      [0], reverse=True)
    for insultid in order:
      print(('<tr><td align="center">%d</td>'+
          '<td align="center">%d</td>'+
          '<td>%s</td><td>%s</td>'+
          '</tr>')%(
          insultid, votes[insultid][0],
           votes[insultid][1],
           stock_actions(insultid)))
    print('</table>')
    print('</blockquote>')

def show_new_insult(insultid, insult):
    "Print HTML that displays a new insult in
     a single-cell table"
    print('<h1>Your New Insult</h1>')
    print('<blockquote>')
    print('<table border="1"><tr><th>ID</
      th><th>New Insult</th>'+
        '<th>Action</th></tr>')
    print(('<tr><td align="center">%d</td>'+
        '<td align="center"><big><b>%s</b>
          </big></td>'+
        '<td>%s</td>'+
      '</tr>')%(
        insultid, insult, new_actions(insultid)))
    print('</table>')
    print('</blockquote>')

def show_generate_insult():
    "Print HTML that displays buttons to
    request insults"
    print('<blockquote>')
    print(new_form_button()+'<br/>'+get
      specific_form_button())
    print('</blockquote>')

def get_insult(insultid):
    "Return a string containing a specific
      insult"
    # set the random number generator seed to
      the value that will
    # cause 'rudeness' to generate the insult
     we seek
    oldstate = random.getstate()
    random.seed(insultid)
```

```python
    # call 'docker' to provide an insult
    insult = rudeness()
    random.setstate(oldstate)
    return insult

def vote_insult(insultid):
    "Vote for the given insult"
    if insultid in votes:
        votes[insultid][0] += 1
    else:
        insult = get_insult(insultid)
        votes[insultid] = [1, insult]
    save_votes()

def delete_insult(insultid):
    "Delete information in the table about the
      given insult"
    if insultid in votes:
        del votes[insultid]
        save_votes()

def handle_op():
    "Interpret the CGI 'op' argument and do
     what it asks"
    query = cgi.FieldStorage()
    op = query.getvalue("op", "")
    insultno = query.getvalue("number", "")
    if op == "new":
        if insultno == "":
            insultid = random.getrandbits(24)
        else:
            insultid = int(insultno)
        insult = get_insult(insultid)
        show_new_insult(insultid, insult)
        show_generate_insult()
    else:
        print('<h1>Try another Insult</h1>')
        show_generate_insult()
        if op == "vote":
            insultid = int(insultno)
            vote_insult(insultid)
        elif op == "delete":
            insultid = int(insultno)
            delete_insult(insultid)
        elif op != "":
            raise VotesError("Bad op - "+op)

def main():
    "Print the whole HTML for the current Web
     page"
    show_top()
    load_votes()
    handle_op()
    show_votes()
    show_bottom()

if __name__ == '__main__':
    main()
```

SOFTWARE RECIPES
Access via Windows Shares

You can gain access to the files on your Raspberry Pi from any other computer on the same network that can use Windows shares.

The 'Server Message Block' (SMB) networking protocol was originally designed by Microsoft in the 1990s to provide Windows computers with a way to share files over a network, giving them the same kind of capability that Unix-based computers had with their Network File System (NFS). Today it's the protocol that your Windows computer uses when you elect to 'share' a folder with users on other computers in your home.

Originally Microsoft devised a complete alternative to the communications methods now (and then) adopted as the basis of the Internet, having its own protocols for transmitting data between computers and for finding their addresses given computer names. These original 'NetBIOS' mechanisms have now been augmented to allow the optional use of the more normal Internet protocols.

An open source version of the SMB protocol and some of the underlying Microsoft protocol was reverse engineered from the Microsoft implementation (that it, it was copied by observing its behaviour) and is now available in Linux as the 'Samba' protocol. It's available both as a server, where local files are provided to remote computers; and as a client, where files on a remote computer are made available locally. In this section we concentrate on the use of Raspberry Pi as an SMB server, although in passing we also look at its use as a client.

Uses of a Raspberry Pi file server
Files shared this way can easily be used by other computers running Linux, Windows or OS X (Apple) operating systems.

A Raspberry Pi running an SMB server is a simple Network Attached Storage (NAS) device. You could, for example, plug a large USB disk into Raspberry Pi which you could then use from the other computers on your home-network.

If you store your music files on this disk you'd be able to play them through these other computers, making a kind of music server.

Alternatively you might just use this storage as a place to copy important files that you want to back up.

If you've set Raspberry Pi to perform some regular function or respond to requests made using its web server, having Raspberry Pi's files available remotely provides a way to pass information on to the other computers on the network. For example, if you write a Python program to control your central heating it can record a log of what it's done, which you'll be able to see in a shared file.

How it works
In order to use an SMB connection Raspberry Pi and the remote computer must both run SMB servers. The connection between the two computers is normally over the Internet, allowing them to be in completely different locations.

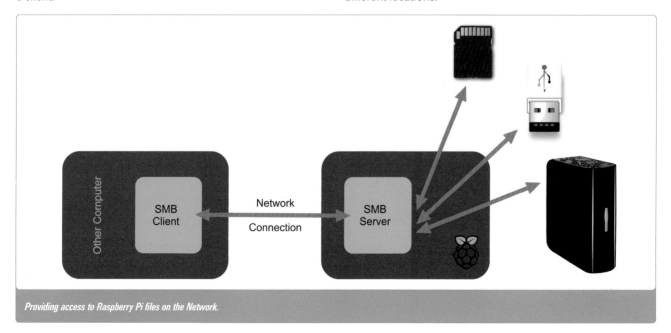

Providing access to Raspberry Pi files on the Network.

Raspberry Pi preparation

Because SMB relies on a network connection you'll have to provide either a wired Ethernet connection or a wireless WiFi connection to your Raspberry Pi first.

Starting the Samba service

You may find that your Raspberry Pi already provides a Samba server. Run the following command to discover whether this is the case:

```
sudo service samba status
```

If a message like 'nmbd is running' and 'smbd is running' is given Raspberry Pi is already set up properly and you needn't do any further preparation. If the message is more like 'unrecognized service' the service will need to be installed from the network. This command should accomplish that:

```
sudo apt-get install samba
```

(It should also set the service to run after every boot and start the service immediately.)

Sharing a directory

Every detail of SMB service is controlled by the file /etc/samba/smb.conf, but changes to this file aren't registered until the Samba service is stopped and started again. Some additions will be needed to this file in order to tell the service which part of the file system is to be shared with others. Adding the following lines to the end of the file will result in a new 'share' being provided out of the directory /public that everyone will be able to use:

```
[public]
    comment = Public files
    path = /public
    valid users = @users
    force group = users
    create group = 0660
    directory mask = 0771
    read only = no
```

To do this you'll have to edit /etc/samba/smb.conf as root, for example with the 'nano' editor:

```
sudo nano /etc/samba/smb.conf
```

(Type the Ctrl-U several times until the end of the file is reached, type in the lines above and then type Ctrl-X, answering 'Y' to the question about saving the modified buffer.)

Before this change is of any real use we'll have to create the directory

```
sudo mkdir /public
```

and make sure that normal users (ie those in the 'users' group) can put things in it.

```
sudo chown -R root:users /public/
sudo chmod -R ug+rwx /public/
```

To test whether users (such as you) can create new things in this directory try this:

```
mkdir /public/new
```

This should execute without an error.

Naturally we could have used any other directory name in place of /public, and it's possible to add extra shares in a similar way.

Restart the SMB server

Run the following to restart the Samba server (which will force it to re-read its configuration file and thus export /public):

```
sudo service samba restart
```

Create a network user

Although any SMB user is allowed to use our new share it may well be that there are no SMB users! Linux distributions don't (often) use the same names and passwords when logging on to Linux that are used for using Samba shares.

Depending on the age of the Linux distribution the command to update the list of SMB users may either be 'smbpasswd' or 'pdbedit'. If you have a standard Raspberry Pi you'll have 'pdbedit'. (Otherwise use 'man' command to discover more about how 'smbpasswd' is used.)

You can list the SMB users that there are (don't be surprised to see nothing printed):

```
sudo pdbedit --list
```

Assuming there's nothing in this list you'll need to create at least one new user. If you're well organized and already have fixed names for your family used by the computer(s) already in your house there's a benefit to creating users with those names. Otherwise we'll simply create an imaginary user 'pifiles' who can use the share. To do this type

```
sudo pdbedit --create --user pifiles
```

This will ask you (twice) for a password for that user (eg 'raspberry'). Don't forget it once you've set it. Not only does this make a new user for SMB but it also creates a new system user with its own home directory (in /home/pifiles).

SMB service information

A client will need to know the name of the SMB server (the computer providing the SMB service). As we said above, however, there's more than one way that a name can be found in SMB, corresponding to the original NetBIOS and the current Internet methods. In addition to at least two other methods there are:

- 'Broadcasting' to all the computers on your home network asking all of them if they have the requested SMB server name.
- The normal Internet way to find an IP address from a computer name.

The section about Networking discusses Internet Protocol addresses (four numbers less than 256 separated by dots) and the way they're named in the Internet. You can find its IP address by typing

```
hostname -I
```

You'll always be able to use this as the name of the SMB server, but it isn't really as convenient as using a real name, so as you probably don't have an Internet name server it would be nice to use the computer's NetBIOS name. It's possible to provide any name you like by updating the /etc/samba/smb.conf file (for example, you could set it to 'lemonpi' by including the line

```
netbios name = lemonpi
```

in the [global] section), but if you haven't done this (most people don't) the system's name is used instead. The system name is the one that

```
hostname
```

reports. On the reference distribution this is 'raspberrypi'.

Naturally, different computers each need their own name, so if you have two Raspberry Pis you'll need to change the name of at least one of them. It's usually possible to change this name yourself just by editing the file /etc/hostname. This would set it to 'blackberrypi':

```
sudo tee blackberrypi -a /etc/hostname
```

If you do change your host name you should be aware that the NetBIOS name for the system should be 15 characters or less for successful operation.

To use the share the client will use either this or its IP address along with information needed to name and authenticate a user.

Surprisingly, although we know we have a user called 'pifiles' its user name can pose a problem. This is because SMB can authenticate users on different computers. If two authenticating computers called, say, 'dee' and 'dum' both have a user 'tweedle' SMB doesn't assume that 'tweedle' at 'dee' is the same person as 'tweedle' at 'dum'. For that reason SMB clients have to know the name of the authenticating machine (the SMB user's name 'domain') as well as the name to use there.

Our user 'pifiles' is really 'pfiles' on our Raspberry Pi computer, and since we're using that computer to do the authentication Raspberry Pi also provides the naming domain.

In summary, you'll need to accumulate these pieces of information about your system:

- Its SMB name (eg 'raspberrypi' or 192.168.104.95).
- Its domain name ('raspberrypi'?).
- The user's name ('pifiles'?).
- The user's password.

SMB client setup

Almost every operating system supports some form of SMB client, including Windows, Linux, OS X and Android.

Windows client setup

Windows has the necessary software installed by default.

In Windows Vista, 7 and 8 there's a 'Network and Sharing Center' in the Control Panel with an option to see advanced sharing settings. You'll need to see those and use them to turn on 'network discovery'.

The first time the Raspberry Pi share is used Windows will try to establish the user's name and password and will remember some or all of them for subsequent uses.

To use our share simply use the SMB server name or IP address preceded by two back-slashes (\\) as a folder name in Explorer. For example:

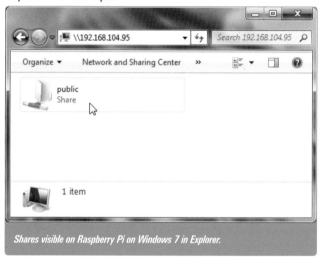

Shares visible on Raspberry Pi on Windows 7 in Explorer.

Once our 'public' share is opened Windows requests authentication details:

Logging on to a share in Windows 7 Explorer.

In older versions of Windows the domain name of the system is assumed to be the same as the computer supporting the share. In later versions, however, the domain used by the host machine is assumed instead. In either case it's safe to quote a name consisting of the domain name followed by the user name after a backslash (\) ('raspberrypi\pifiles' in our example).

If you see a tick box labelled 'Remember my credentials' and tick it the information you type in won't be requested when you return to that share in the future. The name and password you use will be stored and reused instead.

The result should be the contents of the share, which can be used in the same way that other files in Windows can.

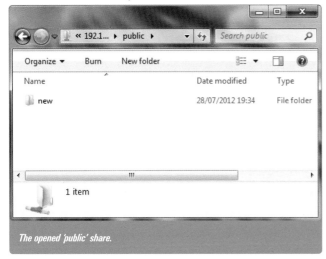

The opened 'public' share.

Linux client setup

The Samba client is normally available through the package manager supported by your Linux distribution. Each distribution has a graphical program that makes it easy to install these packages, but because there are several of these it'll be better to describe the process from a command-line interface.

For example, in Ubuntu, Debian and a few other distributions the following command will list the names of packages whose names include 'samba':

```
aptitude search samba
```

You're liable to find a package with a name like 'samba-client'. This could be installed (if it hasn't already been) using

```
sudo apt-get install samba-client
```

Once installed you can make use of the client either on the command line or in the graphical user interface.

Like the SMB server the SMB client is also controlled by the file /etc/samba/smb.conf and there's currently an infelicity in the version of this file that's used. As we discussed earlier there are a number of ways that an address can be found from an SMB client name. When it needs to, Samba will try each method in turn in some specific order. Unfortunately the normal order results in the 'broadcast to all computers' method being used after an Internet name lookup, even though this is the one that's most likely to work. This will be a problem in a growing number of homes were the Internet address lookup of an unknown name is always redirected to an Internet Service Provider's search engine (in their ISP's effort to gain additional advertising revenue from mistyped web addresses).

To fix this problem you may need to edit this file on the client. In your favourite editor search for a line containing

```
name resolve order =
```

Often it's commented out (with a semi-colon (;) as the first character on its line). If you don't find it search for

```
[global]
```

instead and in either case add a line as follows:

```
name resolve order = bcast lmhosts
    host wins
```

(or any other order that places 'bcast' before 'host').

You may now need to restart the Samba name lookup service. The simple way to do this would be to reboot the computer but

```
sudo service samba restart
```

or

```
sudo service smbd restart
```

may work.

Browsing Samba from a Linux GUI

The graphical user interface can be provided by a number of different desktop window managers in Linux. For example, the Gnome and KDE systems are popular. On Raspberry Pi we use the LXDE system.

All of these have a program that displays a directory which will also be able to show directories held on SMB shares.

Samba use from a Gnome desktop

Most Gnome desktops use the 'nautilus' program to display a directory. You may find that the following package needs to be installed before this program is able to display files available through SMB:

```
sudo apt-get install gvfs-backends
```

(for Linux distributions that use 'apt-get' to install packages).

Use the Go > Location … menu item to allow the specification of an SMB file name:

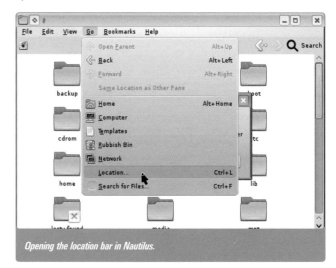

Opening the location bar in Nautilus.

You can then type in the share name for Raspberry Pi which will be 'smb://' followed by its IP address (192.168.104.95 in our example):

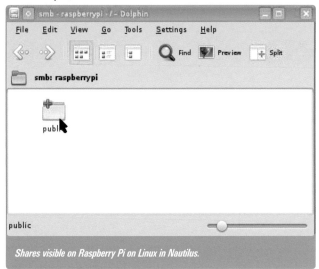

Shares visible on Raspberry Pi on Linux in Nautilus.

Opening our 'public' share will result in the authentication dialogue appearing which will need the normal details:

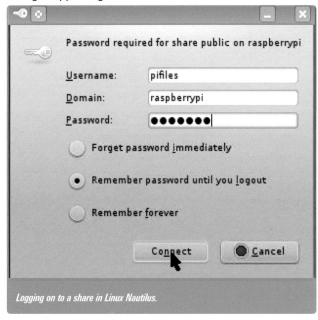

Logging on to a share in Linux Nautilus.

As in Windows 7 you have the choice to remember your logon details (and in this case for how long). The result is a normal directory display which you can use as if the files were local.

The opened 'public' share.

Samba use from a KDE desktop

Most KDE desktops use the 'Dolphin' program to display a directory.

Again you may find that the following package needs to be installed before Dolphin is able to display files available through SMB:

```
sudo apt-get install kdenetwork-filesharing
```

In Dolphin it's possible to type the name of the SMB share into the area where the directory location is usually displayed. Although you don't have to, you can supply the name of the Raspberry Pi user you want to use as part of the share name, typing it before the IP name followed by an at sign (@).

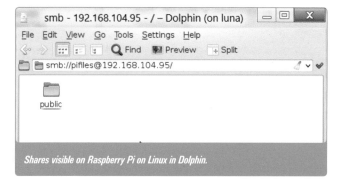

Shares visible on Raspberry Pi on Linux in Dolphin.

Again opening 'public' will open an authentication dialogue:

Logging on to a share in Linux Dolphin.

Again the result will be a normal directory display:

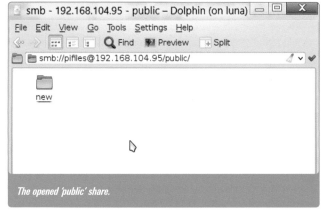

The opened 'public' share.

Samba use from an LXDE desktop

Most LXDE desktops (including the one on Raspberry Pi) use the 'pcmanfm' program. Like the Gnome desktop's Nautilus program you may find that the following package needs to be installed before this program is able to display files available through SMB:

```
sudo apt-get install gvfs-backends
```

If you install this package while running LXDE you'll have to log off and log on again.

Use of this program follows the same pattern as the other file managers. The shares available can be viewed by typing the same name into the browser:

Shares visible on Raspberry Pi on Linux in pcmanfm.

Opening 'public' will again request the usual details:

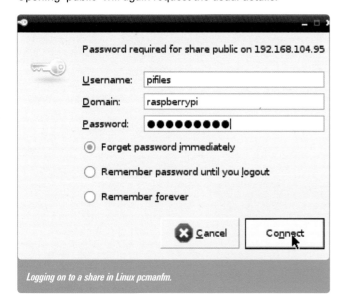

Logging on to a share in Linux pcmanfm.

And the result will appear in a normal pcmanfm directory display:

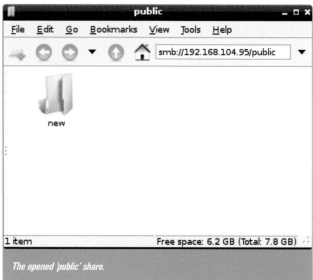

The opened 'public' share.

Mounting an SMB file system in Linux

Linux clients can place the whole of an SMB file system in any directory in such a way that programs cannot easily distinguish files on the remote file system from others outside that directory. For example, we might place the files that Raspberry Pi exports in the 'public' folder in our own subdirectory called /mnt/pipub. To do this we must first ensure that directory's existence using

```
mkdir /mnt/pipub
```

Then we 'mount' the remote filing system at this directory (which is referred to as a 'mount point').

```
sudo mount -t cifs -o user=pifiles //
    raspberrypi/public /mnt/pipub
```

'sudo' executes the rest of the line as the privileged user 'root'. The mount command itself has four parts:

1 The name of the type of file system that's to be used after '-t' ('cifs' stands for Common Internet File System).
2 Options for the mount after '-o'.
3 The location of the file system to mount (//raspberrypi/ public will be understood as a CIFS device name).
4 The directory where the mounted file system will appear.

Running this command will request user 'pifiles's password on the SMB server 'raspberrypi'. You can avoid the password request by including it in the list of options like this (here we assume the password is 'raspberry'.)

```
sudo mount -t cifs -o user=pifiles,password=
    raspberry //raspberrypi/PUBLIC /mnt/pipub
```

If you try this and receive an error (e.g. about an operation already being in progress) try using the IP address of the SMB server instead of its name. Alternatively you could place the names and IP addresses of SMB servers in your home network (assuming they have fixed IP addresses) in the file /etc/hosts.

Examining the contents of /mbnt/pipub now shows the files from Raspberry Pi:

```
$ ls -l /mnt/pipub/
total 0
drwxrwx--x 2 root root 0 Jul 28 19:34 new
$
```

You may notice that all the files are owned by 'root' and belong to the 'root' group. This can be very inconvenient since it'll mean that you may be unable to read or write the files without 'sudo'. To overcome this problem you can force the user ID ('uid') and group ID ('gid') to be ones that you can use. For example, on my remote machine there's a user 'gray' and a group 'users', so I could use the mount command:

```
% sudo mount -o user=pifiles,password=raspberry,
uid=gray,gid=users,forceuid //raspberry/
PUBLIC /mnt/pipub
% ls -l /mnt/pipub
total 0
drwxrwx--x 2 gray users 0 Jul 28 19:34 new
```

(Although the mount command is shown on several lines, it should all be typed on the same one.) You can see that the files are now owned by 'gray' and the 'users' group.

These files will remain available at the mount point until unmounted using the command

```
umount /mnt/pipub
```

Permanent mounts
It's possible to make specific mounts happen automatically whenever Linux boots, for example making Raspberry Pi files always available. Such mounts are listed in the file /etc/fstab.

If you've found that the above mount always works you might like to reuse the details in a single line of this file. The line must contain at least these parts, separated by spaces:

- The name of the CIFS device to mount.
- The directory where it's to be mounted.
- The type of the file system ('cifs').
- The file system options that are to be used.

Normally this is then followed by two zeroes. In our case /etc/fstab should be edited to include a line that looks something like this:

```
//raspberrypi/public /mnt/pipub cifs user=
pifiles,password=raspberry,uid=gray,gid=users,
forceuid 0 0
```

(but all on one line).

You can test this out without rebooting by unmounting /mnt/pipub and then using the command

```
sudo mount -a
```

which will mount any unmounted devices listed in /etc/fstab.

Because everyone on your Linux computer can read the contents of the /etc/fstab file you may feel uneasy about including the password for a mount. Another mount option is available that will make this unnecessary. It gives the name of a file where the user name and password are to be placed. Although it must be readable by the privileged user 'root' this file can be given file access permissions that deny normal users the ability to read it. The mount option is called 'credentials' and it replaces the 'user' and 'password' options so that the line in /etc/fstab might look like this:

```
//raspberrypi/public /mnt/pipub cifs
credentials=/root/picreds,uid=gray,gid=users,
forceuid 0 0
```

The replacement logon details placed in /root/picreds must contain these two lines:

```
username=pifiles
password=raspberry
```

A Raspberry Pi SMB client
Naturally, because Raspberry Pi also uses LXDE the instructions above describe how Raspberry Pi can be used to connect to Samba shared in other SMB servers (whether other Raspberry Pis, Linux machines or Windows machines). Using Samba like this is very useful for many things including:

- Extending the file space available to Raspberry Pi.
- Placing program results in files somewhere that doesn't depend on Raspberry Pi being turned on.
- Accessing files (such as music or films) that are stored on your home PC.

You can also mount remote file systems into Raspberry Pi as above and make them permanent by updating Raspberry Pi's version of /etc/fstab.

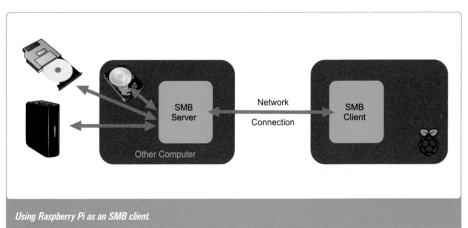

Using Raspberry Pi as an SMB client.

SOFTWARE RECIPES
Access via VNC

The components inside a computer all interact over one or two communications 'buses'. Someone in a small blue-sky research company in Cambridge wondered if this communication could be more remote, replacing the buses with the high-speed Asynchronous Transmission Mode (ATM) network they were working on. Could a computer be 'exploded' into many simpler parts each coexisting on the same network?

That research project led to the production of hundreds of tiny ARM-based computer boards each joined to a fast network (does this sound familiar?) all working together to form one large multimedia computer. In addition to speakers, microphones, cameras and desktop interfaces there was one component, called a 'video tile', which any Unix-based operating system could use as a display, even though it could be anywhere on the ATM network.

One of the main directions of this research was to consider what you'd do if networks were so fast you could use them freely for any purpose, so the simplest possible communication protocol was designed for a video tile called a Remote Frame Buffer (RFB) protocol.

At the time the X protocol (which you'll find in another 'access' section in this book) was available and could have been used, but it was (and still is) full of mechanisms that reduce network use by relying on complicated communication rules and a relatively large amount of intelligence in the display. When the communication link between a computer and the display broke or was closed the protocol had so much state that it was impossible to reinstate the display and the whole session always needed to be started again – usually with a user having to log on and start his computer session again.

The much simpler RFB protocol didn't have this problem. Said to be 'stateless', it was possible to halt the connection at any time but reconnect later and have exactly the same thing displayed, making it very resilient.

Before long all the phones in Olivetti Research Limited (the first of a string of names the company took) were replaced by video tiles augmented with speakers, microphones and a touch screen. In addition to being used as video phones, users could run games and other programs. To the user it looked as if the phone was a private computer, even though the programs were really running on bigger computers elsewhere in the laboratory. In fact the network as a whole gave users a series of not-quite-real virtual computers and the experience was called Virtual Network Computing.

A couple of years before the company was closed to ease its owners' financial problems in America (by then it was owned by AT&T) it took the program used to implement RFB and made it public domain. Since then the program, if not the idea of virtual networking, has been used to provide displays on all kinds of large and small devices and appears on Linux and Windows computers as the VNC suite of programs.

Naturally this program is available on Raspberry Pi, a single-board computer that would have seemed very familiar to the original Olivetti Research team 15 years ago. (I can say this with authority since I was one of them.)

How it works

In VNC one computer acts as a 'client' and the other as a 'server'. The server pretends to be a display to the operating system it runs under, but in fact uses RFB to send its image to the client on a remote computer (which can show the image there).

There are two ways that VNC can be used by Raspberry Pi: one where Raspberry Pi uses VNC to produce its display on another computer; and another where another computer's display is shown on the television connected to Raspberry Pi. Both of these are useful but this section is going to concentrate on the former, because it's another way to use your Raspberry Pi from somewhere else.

Imagine that you're using your Raspberry Pi to control a weather vane in your loft, or your central heating from the airing cupboard in your house. You won't want to attach a television or monitor to it in either of these cases. Using VNC you can still use Raspberry Pi's graphical user interface, but from a different computer in your house, or even your mobile phone.

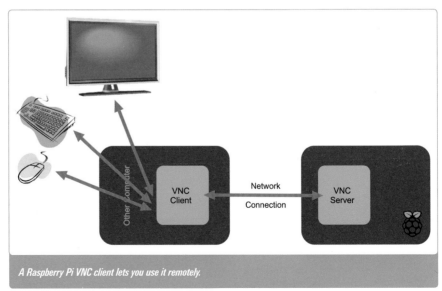

A Raspberry Pi VNC client lets you use it remotely.

Raspberry Pi preparation

Because VNC relies on a network connection you'll have to provide either a wired Ethernet connection or a wireless WiFi connection to your Raspberry Pi first.

In order to use VNC, Raspberry Pi and the computer that you're using must both run VNC servers. Typing this command

```
sudo apt-get install vnc-server
```

will install the server software needed on Raspberry Pi.

Running the VNC server

On Raspberry Pi any number of GUI interfaces can be started for VNC clients – all completely separate from each other. Each VNC session is distinguished from the others by a small number which can be provided when the server is started. As well as being separate from each other (ie each can be used by a different user without interfering with each other), each of these sessions is separate from the X session GUI that Raspberry Pi sometimes starts. In fact Raspberry Pi doesn't have to run X on the TV's display at all.

It's simple to run the VNC server. In this example we make VNC session number zero, which is referred to as ':0'. Just type

```
vncserver :0
```

You could make other sessions with different session numbers in the obvious way.

The first time this is run you'll have to provide a password for the client to use. The password cannot be too short (eg 'pi'):

```
pi@raspberrypi ~ $ vncserver :0

You will require a password to access your
desktops.

Password:
Password too short
pi@raspberrypi ~ $
```

and if it's too long (eg 'raspberry') only the first eight characters will be used:

```
pi@raspberrypi ~ $ vncserver :0

You will require a password to access your
desktops.

Password:
Warning: password truncated to the length of 8.
Verify:
Would you like to enter a view-only password
(y/n)? n
New 'X' desktop is raspberrypi:0

Starting applications specified in /home/pi
/.vnc/xstartup
Log file is /home/pi/.vnc/raspberrypi:0.log

pi@raspberrypi ~ $
```

Once you see 'New 'X' desktop is raspberrypi:0' you've created the server.

The server will give VNC clients access to the new GUI until you delete it. You can delete the server using the command

```
vncserver -kill :0
```

Making the VNC server always available

When you reboot Raspberry Pi you'll lose the server too, so either you'll have to recreate it on each occasion or you should run the service automatically each time Raspberry Pi boots.

We briefly described how Linux reads and executes Bash scripts in the directory /etc/init.d when it boots in the section about Linux.

Unfortunately Debian doesn't supply a script for installing a VNC server in this way, but we can add one of our own. To do this create a new script running as the privileged 'root' user with your chosen editor

```
sudo editor /etc/init.d/vncserver
```

Then enter this shell script:

```
#!/bin/sh -e
#
# start/stop the vncserver daemon
#
### BEGIN INIT INFO
# Provides:            vncserver
# Required-Start:      $network $remote_fs
# Required-Stop:       $network $remote_fs
# Default-Start:       2 3 4 5
# Default-Stop:        0 1 6
# Short-Description: Start vncserver
# Description:         vncserver uses the Remote
                       Fame Buffer Protocol
#                      allowing connection from
                       a VNC client.
### END INIT INFO

USERID=pi
NAME=vncserver
DESC="VNC session server"
VNC_DISPLAY=:0
VNC_SERVER=/usr/bin/vncserver

. /lib/lsb/init-functions

FULLVNC_DISPLAY=`hostname`$VNC_DISPLAY
PIDFILE=/home/$USERID/.vnc/$FULLVNC_DISPLAY.pid

# Carry out specific functions when asked to by
  the system
case "$1" in
  start)
          log_daemon_msg "Starting $DESC "
          if [ -e "$PIDFILE" ]; then
          log_progress_msg "$NAME already running"
          log_end_msg 0
```

```
      else
        log_progress_msg $NAME
        su $USERID -c "$VNC_SERVER $VNC_
         DISPLAY &>/dev/null"
        log_end_msg $?
      fi
      ;;
  stop)
      log_daemon_msg "Stopping $DESC "
      log_progress_msg $NAME
      if [ -e "$PIDFILE" ] > /dev/null; then
        su $USERID -c "$VNC_SERVER -kill
         $VNC_DISPLAY &>/dev/null"
        log_end_msg $?
      else
        log_progress_msg "not running"
        log_end_msg 0
      fi
      ;;
  *)
    echo "Usage: /etc/init.d/$NAME {start|stop}"
    exit 1
    ;;
esac

  exit 0
```

and make sure it can be executed:

```
sudo chmod +x /etc/init.d/vncserver
```

This will respond to the 'start' and 'stop' requests expected by the scripts in this directory.

Now insert the service into the start-up system so that it runs on boot:

```
sudo insserv vncserver
```

and test that it works by starting the service

```
root@raspberrypi:/home/pi# sudo service
 vncserver start
[ ok ] Starting VNC session server : vncserver.
```

and stopping it.

```
root@raspberrypi:/home/pi# sudo service
 vncserver stop
[ ok ] Stopping VNC session server : vncserver.
```

VNC service name
The section about Networking discusses Internet Protocol addresses (four numbers less than 256 separated by dots) and the way they're named on the Internet. Again you can find its IP address using

```
hostname -I
```

The client will use this IP name followed by a colon (:) and the number of the VNC session to make the name of the

VNC session remotely. So if Raspberry Pi's IP address is 192.168.1.95 and we've created session :0 the remote name of the session will be

```
192.168.1.95:0
```

Security
The password exchange that happens when a client logs on isn't secure – other computers on the network can see what it is. An SSH tunnel (briefly mentioned in the section about SSH) is a good way to protect the password if the remote computer is on the Internet.

Not very fast
As mentioned above, RFB was designed to be simple, not fast. Over a slow network other ways to share the desktop might be faster. In any event it's very likely that very rapid changes to the desktop won't be adequately reflected on the remote display. Video, for example, won't travel well.

Also, the special GPU hardware available in Raspberry Pi to speed up the display isn't used by the VNC server and so provides no speed benefit.

Because a WiFi network connection is (almost always) slower than a directly connected Ethernet you may find that VNC is more responsive over Ethernet.

VNC client setup
Almost every operating system supports some form of VNC client, including Windows, Linux, Mac OS and Android. If the instructions here aren't sufficient you can almost certainly find what you need using your favourite search engine.

Windows client setup
There are quite a few Windows clients that will connect to a VNC server. Without being specific, you'll need to find the client program on the Web, download the installation program and run it to obtain one. One popular open source, and free, client is called TightVNC, which is available from

```
http://www.tightvnc.com/download.php
```

At its web page select the download appropriate for your version of Windows (it'll be either a 32- or 64-bit version – the system properties, displayed using the Windows button and the Break or Pause button at the same time, will tell

Download the TightVNC installation program.

you which). Click on it, save the resulting file and run it. It will announce itself and then, once you have clicked on 'Next' ask you to agree to the GNU Public License. 'Next' will take you on to a page asking what you want to install, another 'Next' takes to to a summary of what the installer is about to do.

By default it will install both a VNC client and a VNC server on your computer. If you want to install only the viewer use the advanced installation option. Finally you'll be asked whether you want to go ahead with the install and you should

By default both the client and the server are installed.

click on the 'Finish' button. Windows will check that you intended to perform the installation and then go ahead.

Now it's installed you'll be able to use the program 'TightVNC' in the normal way using the menu Start > Programs > TightVNC > TightVNC Viewer. It will ask for the name of the VNC session you want to connect to. For example, you might type:

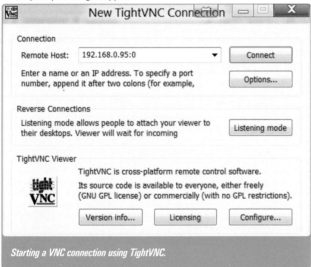
Starting a VNC connection using TightVNC.

You will then be asked for the password, which you must type in correctly. If successful the VNC session running on Raspberry Pi will be displayed.

Linux client setup
Like Windows, there are several VNC clients available for Linux. Normally they'll be available through the package manager supported by your Linux distribution. Each distribution has a graphical program that makes it easy to install these packages, but because there are several of these it'll be better to describe the process from a command-line interface. Naturally, though, these command lines must be

run from a Linux that supports a GUI (because it'll be used to display the VNC session). For example, in Ubuntu, Debian, and a few other distributions the following command will list the names of packages whose names include 'vnc':

```
aptitude search vnc
```

You're liable to find a package with a name like 'vnc-viewer'. This could be installed (if it hasn't been already) using

```
sudo apt-get install vnc-viewer
```

Once installed the name of the client command is usually 'vncviewer' and it can be run by typing its name followed by the name of the VNC session. For example, if we have a Raspberry Pi session called 192.168.1.95:0 it can be displayed using

```
vncviewer 192.168.1.95:0
```

This will normally request the password first:

```
pi@raspberrypi ~ $ vncviewer 192.168.1.95:0
Connected to RFB server, using protocol
 version 3.8
Enabling TightVNC protocol extensions
Performing standard VNC authentication
Password:
Authentication successful
Desktop name "pi's X desktop (raspberrypi:0)"
VNC server default format:
  32 bits per pixel.
  Least significant byte first in each pixel.
  True colour: max red 255 green 255 blue 255,
   shift red 16 green 8 blue 0
Using default colormap which is TrueColor.
 Pixel format:
  32 bits per pixel.
  Least significant byte first in each pixel.
  True colour: max red 255 green 255 blue 255,
   shift red 16 green 8 blue 0
```

Again, the result should be a window showing Raspberry Pi's GUI.

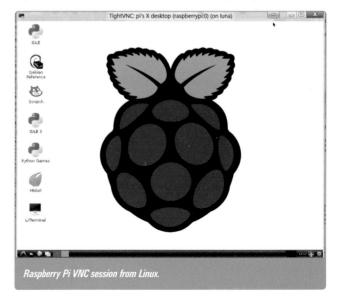
Raspberry Pi VNC session from Linux.

SOFTWARE RECIPES
Access as a Windows Remote Desktop

IT released the X Windows system in 1984, and, long before Windows 95 – or even Windows 3 – was released, provided a well thought-out means to access a desktop computer remotely. Microsoft, a company whose income depends on the number of its own operating systems providing desktops to users, unsurprisingly didn't adopt this de facto software. Instead it licensed software from Citrix that became the Windows Terminal Services (now called Remote Desktop Services). Like X this system uses a server that provides access to a computer and clients that take advantage of that access. In this case, though, the network protocol used was Microsoft's own Remote Desktop Protocol (RDP).

Most new versions of Windows have had an associated version of RDP, including both client and service software, which advances its version alongside the main operating system (meaning that older versions of Windows cannot always use newer versions remotely). Different Windows licences permit a different number of remote connections and there's normally a means to pay for an upgrade that can increase the number of potential users.

Windows Remote Desktop is better integrated into Windows than non-Microsoft versions of X and is faster than VNC. Furthermore the licensing arrangements for running the RDP client on Windows are much better defined than those for running the relevant X or VNC software (which may be against the terms of a Windows licence).

Non-Microsoft implementations of RDP have been made that provide both clients and servers for other operating systems. The regular advent of a newer version of the protocol means that older client implementations may not connect to new versions of the server, but otherwise the system is quite usable on other systems, including Linux.

You can therefore gain access to your Raspberry Pi from other computers which can use an RDP client (notably Windows computers).

How it works

As with VNC, one computer acts as a 'client' and the other as a 'server'. Again the server pretends to be a display to the operating system it runs under, but in fact uses RDP to send its image to the client on a remote computer (which can show the image there). It also does a similar job with mice, keyboards and sound cards – potentially allowing them all to be used remotely by the client.

There are two ways that RDP can be used by Raspberry Pi: one where Raspberry Pi uses RDP to produce its display on another computer; and another where another computer's display is shown on the screen connected to Raspberry Pi. Again both of these are useful but this section is going to concentrate on the former.

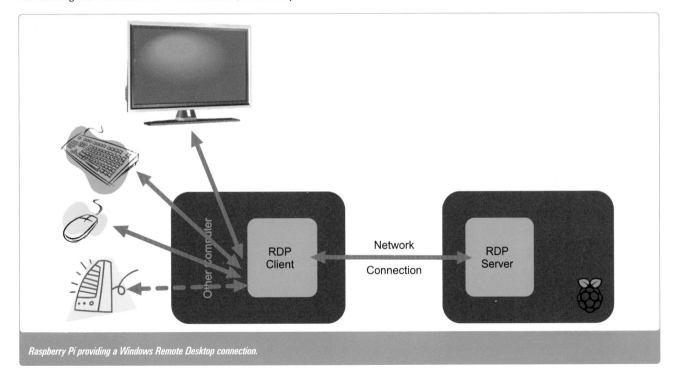

Raspberry Pi providing a Windows Remote Desktop connection.

Raspberry Pi preparation

The 'xrdp' program uses a VNC server to provide the virtual desktop it needs to provide RDP access, so we need to install both xrdp and a VNC server. We do this using

```
sudo apt-get install xrdp vnc-server
```

Although RDP provides the features necessary to carry audio from Raspberry Pi back to the remote computer, that feature isn't directly supported by xrdp.

VNC is used to provide the authentication the client must use to prove it can use Raspberry Pi, so before running xrdp for the first time it's wise to set up a VNC password. The section about VNC described how this could be done.

RDP Service Name

The section about Networking discusses Internet Protocol addresses (four numbers less than 256 separated by dots) and the way they're named on the Internet. You'll probably need to use

```
hostname -I
```

The client will use this IP name to identify the RDP session for itself. So if Raspberry Pi's IP address is 192.168.1.95, that will be the name of the RDP session.

Windows RDP client setup

In most versions of Windows no setup is necessary. The Remote Desktop Connection program is already installed and available through the menu at Start > Programs > Accessories > Remote Desktop Connection.

Linux RDP client setup

On Linux the RDP client is called 'rdesktop'. The installation package for this will be available through the package manager supported by your Linux distribution. Each distribution has a graphical program that makes it easy to install these packages, but, because there are several of these it'll be better to describe the process from a command-line interface. Naturally, though, these command lines must be run from a Linux that supports a GUI (because it'll be used to display the VNC session). This is the command line that will install rdesktop:

```
sudo apt-get install rdesktop
```

Once installed the name of the client command is usually 'vncviewer' and it can be run by typing its name followed by the name of the RDP session. For example, if we have a Raspberry Pi session called 192.168.1.95 it can be displayed using

```
rdesktop 192.168.1.95
```

This will normally open a new window to request the password.

If it is not selected already choose the module 'sesman-Xvnc' and then replace the username with 'pi' and the password with the password you have set for that user

when you set up the VNC server. The result should be a window showing Raspberry Pi's GUI.

Linux client graphical setup

Linux supports a small number of different desktops (such as KDE, Gnome and LXDE, the one that Raspberry Pi uses). Many of these will have helper applications that allow the GUI to be used to launch RDP sessions. It may have a different package name in different distributions. For example, the one for a Gnome desktop on the Ubuntu Linux distribution can be installed using

```
sudo apt-get install gnome-rdp
```

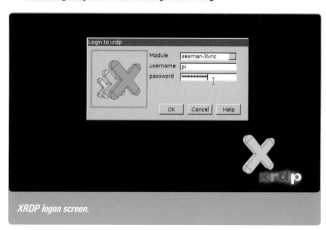

XRDP logon screen.

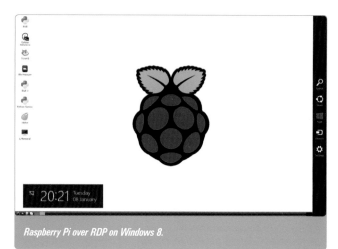

Raspberry Pi over RDP on Windows 8.

05.

Hardware recipes

HARDWARE RECIPES
Using USB Memory Sticks and USB Drives

You can add extra storage to your Raspberry Pi by attaching USB drives of all types. There's no reason why your Raspberry Pi shouldn't have access to many terabytes of storage or access data on CDs, DVDs and Blu-ray disks.

Extra storage can be used for all kinds of purposes including

■ Containing large audio or video collections.
■ Providing a portable source of plug-in private data.
■ Providing network attached storage for other computers in your home to use.
■ Backing up Raspberry Pi data.
■ Providing space for extra programs.

Raspberry Pi has to use an SD card because its initial boot program must be kept there. This has quite a lot of storage (up to about 32GB can be provided on the same SD card). The advantages of using an external source of data on USB are that

■ It can be plugged in, taken out and swapped without having to reboot Raspberry Pi.
■ It can contain a lot more data than 32GB.

What to use
The three types of storage device of greatest interest are

■ USB 'sticks' (sometimes called 'thumb' drives).
■ USB disk drives.
■ USB optical drives.

Thumb drives
Thumb drives have a similar cost and range of sizes to SD cards. In fact it can be quite cost-effective to buy a £1 or £2 USB micro-SD card adapter to make a large thumb drive from.

A 2Gb, 8Gb and 32Gb thumb drive (actually a 32GB micro-SD card in a minuscule USB adapter).

Thumb drives are simple to use on Raspberry Pi. If you have a spare USB socket all you need to do is plug it in directly to the board.

If you don't have a spare USB socket you can create one by displacing whatever you have in a USB socket, plugging in a USB hub instead and plugging the displaced object into the hub (which should have extra USB sockets for other things).

If you're running the graphical user interface you should find that the desktop spots the new USB drive and asks you what to do with it:

LXDE asking what you want to do with a new drive.

Disk drives
Physically two sizes of disk are provided. The smaller ones are 2½in and the larger are 3½in. Naturally the maximum amount of data also varies with the size of the disk, but even 2½in drives can currently contain a terabyte (a million million bytes). With full DVD films using in the order of a thousand million bytes this is still enough room for hundreds of films, and films are by far the largest objects that you'll probably want to store.

Although not a hard-and-fast rule, it's roughly true that 2½in drives are normally powered from their USB cable while 3½in drives normally rely on their own power

Two 2in USB drives, USB 2.0 and USB 3.0.

3in USB drive with its own power supply.

A powered USB hub has its own power supply.

supply. This is quite relevant to Raspberry Pi because, as we described in the section about putting your Raspberry Pi together, its USB socket cannot provide very much power.

A disadvantage of a 3½ inch drive is that its power supply can fail well before the disc which, because they often have non-standard connectors, can sometimes be difficult to replace.

If you plug a drive into Raspberry Pi that takes its power from the USB cable then it's best to plug it in via a 'powered' USB hub.

One of the cheapest ways to obtain a hard drive is to reuse one from an old PC or laptop using a drive caddy. 2½in laptop drive caddies are currently available for as little as £3 from online retailers, but they take their power from their USB connection, so you may need a powered USB hub to drive them (which might cost £15). Caddies for the larger PC (3½in) drives require their own power supply and so are more expensive at about £10.

Optical drives

Optical drives include CD, DVD and Blu-ray disk drives. They can read data from those kinds of disk and, with the right drive and optical medium, can sometimes also write to them.

In addition, music and video can normally be played from CDs and DVDs. There are issues regarding playing films held on Blu-ray disk, however, that mean Raspberry Pi cannot play these (this is explained in more detail in the section about the XBMC media player).

Like newly inserted 'thumb' drives, Raspberry Pi's graphical desktop interface will prompt you for a way to handle different kinds of optical media when they're connected.

Like disk drives, optical drives can also be salvaged from old computer equipment and used in cheap purpose-built caddies, which will normally require their own power supply.

Because they almost all come with their own power supply optical drives don't need to be connected via a powered USB hub. However, the few that do take their power from their USB connection will need to use one.

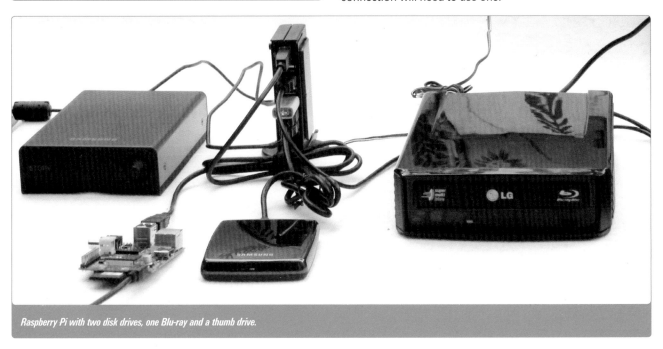
Raspberry Pi with two disk drives, one Blu-ray and a thumb drive.

When they are spotted by Raspberry Pi's graphical interface (LXDE) they are mounted automatically in the directory '/media' and display in the 'places' section of the file manager:

The drives mounted in /media.

Multiple drives

If you plan to use a collection of media for different purposes you'll need a USB hub, and you'll gain the most flexibility by selecting a powered one. If you open a terminal window you can type the command

```
lsusb
```

(**list usb**) to see which USB devices have been found.

```
pi@raspberrypi ~ $ lsusb
Bus 001 Device 001: ID 1d6b:0002 Linux Foundation 2.0 root hub
Bus 001 Device 002: ID 0424:9512 Standard Microsystems Corp.
Bus 001 Device 003: ID 0424:ec00 Standard Microsystems Corp.
Bus 001 Device 009: ID 0409:005a NEC Corp. HighSpeed Hub
Bus 001 Device 005: ID 20f4:648b TRENDnet TEW-648UBM 802.11n 150Mbps Micro Wireless N Adapter
 [Realtek RTL8188CUS]
Bus 001 Device 014: ID 2008:2018
Bus 001 Device 011: ID 0409:005a NEC Corp. HighSpeed Hub
Bus 001 Device 012: ID 04e8:1f09 Samsung Electronics Co., Ltd
Bus 001 Device 016: ID 04e8:5f06 Samsung Electronics Co., Ltd
Bus 001 Device 015: ID 152d:2336 JMicron Technology Corp. / JMicron USA Technology Corp. Hard Disk Drive
```

and use

```
df
```

(**d**isk **f**ree) to see how much storage space is now available:

```
pi@raspberrypi ~ $ df
```

Filesystem	1K-blocks	Used	Available	Use%	Mounted on
rootfs	7579104	5469380	1724876	77%	/
/dev/root	7579104	5469380	1724876	77%	/
devtmpfs	94488	0	94488	0%	/dev
tmpfs	18912	520	18392	3%	/run
tmpfs	5120	0	5120	0%	/run/lock
tmpfs	37820	0	37820	0%	/run/shm
/dev/mmcblk0p1	57288	35232	22056	62%	/boot
/dev/sdb1	732572000	411574544	320997456	57%	/media/SAMSUNG_Videos
/dev/sdc1	1464778368	712982816	751795552	49%	/media/BIG_DISC
/dev/sda1	1993408	15616	1977792	1%	/media/SILVER_2G_2
/dev/sr0	702610	702610	0	100%	/media/OWV_76

(The final entries created in /media are the 2½in and 3½in USB drives, the thumb drive and the Blu-ray drive containing a DVD respectively.)

HARDWARE RECIPES
Using USB WiFi Adapters

You can let your Raspberry Pi use the Internet without needing an Ethernet cable by installing a USB WiFi adapter. This will be particularly useful if you plan to deploy Raspberry Pi in a remote or difficult-to-access location, such as an airing cupboard, a garage or a loft. (Naturally you'll need to ensure that there's a WiFi signal wherever this is.)

USB on Raspberry Pi
Before you consider adding a WiFi adapter ('dongle') there are a couple of things you need to know about the USB interface on Raspberry Pi. Raspberry Pi comes in two flavours: Model A, and Model B.

The single USB interface on the Model A is provided directly by the chip at the heart of the board (the BCM 2835) and, as such, provides very little power.

WiFi dongles typically receive all of their power through their USB connection, and the power available on a Model A is very unlikely to provide as much as is necessary. In this case you shouldn't plug the WiFi dongle directly into the Raspberry Pi. Instead you should obtain a powered USB switch and plug it into that, plugging the WiFi dongle into the switch. Powered USB switches have their own power supply which they can use to supply the devices plugged into it. Naturally it's important only to operate such a switch when it has its power supply connected and turned on. You may find that some WiFi dongles take so much power that a powered USB switch is also required on the Model B too, but most don't.

The symptoms of plugging in a USB device that's taking 'too much' power from Raspberry Pi include erratic behaviour, and sometimes failure to boot.

WiFi adapters and Linux
It may not have escaped your notice that home computers typically run a Windows rather than a Linux operating system. A side-effect of this unfortunate predilection is that manufacturers who develop hardware for home computers can make sufficient profit even when they restrict their customers to the Windows operating system. Often they won't bother creating drivers and other supporting software for other operating systems such as Linux. Although there's a thriving community of programmers who constantly attempt to provide such material it isn't unusual for newer hardware to have no support in Linux. Unfortunately WiFi adapters are susceptible to this problem and you cannot be 100% sure that an arbitrary new WiFi dongle will be usable.

This together with potential power problems, and the fact that some drivers aren't intended for use on an ARM, means that you should be cautious when choosing a WiFi dongle for Raspberry Pi.

The Raspberry Pi community maintains a list of working (and non-working) hardware at

```
http://elinux.org/RPi_VerifiedPeripherals#
Working_USB_Wifi_Adapters
```

In addition to listing the dongles that have been found to work this page also includes installation notes about each.

WiFi adapter firmware
Most WiFi dongles are, in point of fact, miniature computers that execute a fixed program to function. The program that they run is generically termed 'firmware', and it's part of the responsibility of any operating system that supports the dongle to ensure that it's booted with the correct firmware before it's used.

It's often the case that the original manufacturers of the dongle are quite sensitive about allowing the source of their firmware to be seen by their competitors, so the firmware isn't 'open' to all. This provides ethical problems to some Linux distributions, particularly if they don't allow 'non-free' software to be included.

The solution to their dilemma is often to support a separate source of packages that can optionally be either included or not included by the user. In the Debian distribution that forms the core of the Raspberry Pi reference you'll find the word 'non-free' included in the package source specification /etc/apt/sources.list indicating that such firmware packages can be installed.

How it works
A WiFi network delivers a Media Access (MAC) link layer service provided by one of the IEEE 802.11 protocols. These protocols use one of a number of wireless communication channels, each operating at a specific range of radio frequencies. Computers that use one or more of these channels together form a grouping identified by a common Service Set Identifier (SSID), which is just a name with up to 32 case-sensitive letters in it. It's quite possible for a house to support more than one WiFi network, each identified by its own SSID.

Because adapters working on one channel cannot really hear traffic on other channels you could compare the adapters working together in an SSID group to a small crowd of people talking to each other in a big room. When one person shouts the others can hear them, but when two people shout at the same time the resulting 'collision' means that no one can understand what either of them said. To try to avoid these collisions the people employ a politeness protocol. Before a long speech a talker will first announce that he's ready to send a number of words to someone else. That person will quickly reply that he's clear to send that number of words. Everyone else, having heard this initial exchange, will keep quiet for the duration of the talker's speech so as not to interfere. Of course, sometimes people might miss the exchange because they were too far away to hear and might collide anyway, but this shouldn't happen too often. When it does, though, it's possible that the speech will simply get lost.

Although other configurations are possible, it's normal for each SSID to be supported by a single WiFi router which will usually have some connection to another MAC link layer network, such as Ethernet inside your house or 'ADSL' outside, given to you by your Internet Service Provider (ISP).

One area that's always discussed with IEEE 802.11 is its rather poor track record in security. Imagine walking past a window in the room of polite shouters described above. One, perhaps, would have the task of communicating your password to another in the group on its way to the Internet. As a passer-by you may easily hear this and decide to use that information yourself later. This lack of confidentiality is just as obvious in a WiFi network as it is in this imaginary room, so early in the development of the WiFi protocols a method of enciphering the data was devised called Wired Equivalent Privacy (WEP), whereby a special, hopefully un-guessable, code was used to send the data. Sadly that code was cracked and WEP actually provides very little confidentiality now. It's not difficult to find programs that'll simply decode any data collected. New replacement encipherment mechanisms were devised called WiFi Protected Access (WPA) and WPA2, which are more secure, although some (less major) vulnerabilities have been found in them too.

Preparation

In order to get WiFi working on Raspberry Pi we need to accomplish these things:

- Identify the chip used by your adapter.
- Find and install the firmware it requires.
- Install the Linux WiFi driver.
- Declare a new WiFi network interface.

Unfortunately it's not possible to say exactly what the initial step will involve because it varies from one WiFi adapter to another. However, it's possible to illustrate the process with a specific example – a TrendNET TEW-648UBM purchased for £12 on the Web.

Identify your WiFi chip

This is fairly straightforward except for one complication. In principle all you need do is plug the chip into the USB socket and use the command

```
lsusb
```

Unfortunately Linux will try to find drivers for the chip immediately and, in some cases, the wrong one is used and this will crash Linux. If you're unlucky enough to be in this position here are two main suggestions: (1) look your adapter up on the Internet (for example at http://linux-wless.passys. nl/); or (2) find an x86 Linux workstation and use 'lsusb' there.

If you do succeed in using 'lsusb' you'll expect to get output that identifies each of the USB devices plugged in:

```
pi@raspberrypi ~ $ lsusb
Bus 001 Device 001: ID 1d6b:0002 Linux
    Foundation 2.0 root hub
Bus 001 Device 002: ID 0424:9512 Standard
    Microsystems Corp
Bus 001 Device 003: ID 0424:ec00 Standard
    Microsystems Corp
```

```
Bus 001 Device 004: ID 20f4:648b TRENDnet TEW-
    648UBM 802.11n 150Mbps Micro Wireless N
    Adapter [Realtek RTL8188CUS]
```

In the above example the WiFi adapter is made by TRENDnet and we can see that it's identified as '20f4:648b'. It's possible to use this designation on the Web to discover the likely WiFi chip (simply typing it into Google will give a good indication), but in this case the chip is already identified in square brackets: it is a 'Realtek RTL 8188CUS'.

Find the adapter firmware

Now that we know what chip we need a driver for we can search our distribution's packages for it. The idea here is to assume that whatever the firmware package is, its name will incorporate the name of the chip.

As we noted above, it's quite likely that this package, if it exists, will be in a 'non-free' section of Debian repository, so it's important that these are being used in /etc/apt/sources.list.

For example, to search for the 'Realtek RTL 8188CUS' chip you might type

```
aptitude search rtl8188cus
```

and find nothing. Hoping that there might be a package for Realtek firmware in general you might continue:

```
pi@raspberrypi ~ $ aptitude search realtek
i   firmware-realtek
- Binary firmware for Realtek wired and
wireless network adapters
```

which is much more promising.

Assuming you've found such a package you should install it:

```
sudo apt-get install firmware-realtek
```

and then try it out. First unplug the adapter if it's already plugged in. Normally the package will have provided a kernel module to supply the relevant driver and once the USB dongle is inserted that module will be loaded. This command displays the list of modules that are in use:

```
lsmod
```

Note which modules there are and then plug in the USB WiFi dongle (it should be safe now) and repeat the command:

```
pi@raspberrypi ~ $ lsmod
Module            Size   Used by
snd_bcm2835       21485  0
snd_pcm           82208  1 snd_bcm2835
snd_page_alloc    5383   1 snd_pcm
snd_seq           59808  0
snd_seq_device    6920   1 snd_seq
snd_timer         21905  2 snd_seq,snd_pcm
snd               57668  5 snd_timer,snd_seq_
                             device,snd_seq,snd_
                             pcm,snd_bcm2835
8192cu            512098 0
```

In this listing the new module that appears is called '8192cu', and the fact that it appeared is a good indication that the right firmware package has been found.

As a further indication of the success or otherwise of the driver look at the kernel messages that were produced when the dongle was plugged in using the 'dmesg' command. It should mention the driver. For example:

```
[   12.897846] usbcore: registered new
    interface driver rtl8192cu
[   58.805352] usb 1-1.3: new high speed USB
    device number 4 using dwc_otg
[   58.907496] usb 1-1.3: New USB device
    found, idVendor=20f4, idProduct=648b
[   58.907540] usb 1-1.3: New USB device
    strings: Mfr=1, Product=2, SerialNumber=3
[   58.907562] usb 1-1.3: Product: 802.11n
    WLAN Adapter
[   58.907578] usb 1-1.3: Manufacturer: Realtek
[   58.907593] usb 1-1.3: SerialNumber:
    00e04c000001
```

which mentions the name of the chip in the first line.

Finally you should be able to observe that you have a new wireless LAN device by using the command

```
ifconfig -a
```

It will normally be called 'wlan0'.

Now you have confidence that the module '8192cu' is doing a good job you should ensure that it's automatically loaded every time Linux boots. To do this edit the file /etc/modules to include a single line containing the module name. You must edit this file as the privileged 'root' user, so you might type

```
sudo editor /etc/modules
```

(Go to the bottom of the file and type in the name of your module on a line by itself.)

Install the WiFi software
The programs to use WiFi are installed by the command

```
sudo apt-get install wireless-tools
```

The two main kinds of security supported by WiFi were mentioned above: the discredited WEP and the newer WPA. If you're going to use the latter you'll also need this software:

```
sudo apt-get install wpasupplicant
```

It's probably just as well to install the software anyway, in case you move to the superior WPA later.

Declare the WiFi network interface
You'll need to know these pieces of information before you continue:

■ Whether you're using WEP or WPA.
■ The SSID you want to use.
■ The WEP or WPA password the SSID uses.

Now you have a partly working USB adapter you may be able to use it to find the names of the SSIDs available. Assuming it was named 'wlan0', type:

```
pi@raspberrypi ~ $ iwlist wlan0 scan
wlan0   Scan completed :
        Cell 01 - Address: 00:24:A5:BC:B9:A4
                  ESSID:"Office"
                  Protocol:IEEE 802.11bg
                  Mode:Master
                  Frequency:2.412 GHz (Channel 1)
                  Encryption key:on
                  Bit Rates:54 Mb/s
                  IE: Unknown: DD5C0050F2041...
                  Quality=6/100
                   Signal level=44/100
        Cell 02 - Address: 00:12:17:DD:AE:47
                  ESSID:"home"
                  Protocol:IEEE 802.11bg
                  Mode:Master
                  Frequency:2.462 GHz (Channel 11)
                  Encryption key:on
                  Bit Rates:54 Mb/s
                  Quality=100/100
                   Signal level=100/100
```

In this case two SSIDs are available, one called 'Office' and another called 'home'. You should choose the one you wish to use and use exactly the same characters to name your SSID (including whether some of them are upper case). If in doubt simply choose the one with the biggest signal level.

You, or whoever set up your WiFi, will have chosen the password for WEP and/or WPA and may have selected which of WEP or WPA should be used. There's a little difference in setting the two up.

The file containing definitions for Raspberry Pi's network interfaces is /etc/network/interfaces, which you should edit as root using 'sudo'. For example using

```
sudo editor /etc/network/interfaces
```

Declaring a WPA interface
Assuming that 'ifconfig -a' revealed that our new WiFi interface is called wlan0, add these lines to the end of the interfaces file:

```
auto wlan0
iface wlan0 inet dhcp
wpa-conf /etc/wpa.conf
```

which will mean that WPA is controlled by a file we'll call /etc/wpa.conf. If you've elected to join the WiFi network with SSID 'Office' and have found that your WPA password is 'EniGma' you would edit this file (again using 'sudo') so that it contained:

```
network={
ssid="Office"
proto=RSN
key_mgmt=WPA-PSK
pairwise=CCMP TKIP
group=CCMP TKIP
psk="EniGma"
}
```

Declaring a WEP interface

Again, assuming that 'ifconfig -a' revealed that our new WiFi interface is called wlan0 and that we're using the SSID 'Office' with a WEP password '496e5365637565' (WEP passwords must be given in hexadecimal and are normally either 64 or 128 bits long), add these lines to the end of the interfaces file:

```
allow-hotplug wlan0

auto wlan0

iface wlan0 inet dhcp
wireless-essid Office
wireless-key 496e5365637565
```

If your WEP password was set using eight or sixteen characters you can convert it to hexadecimal by replacing each of the characters by its two hexadecimal digit equivalent as displayed by the command

```
man ascii
```

For example, if you've set 'InSecure' as your password it would be converted to '496e5365637565'.

Using the WiFi adapter

Once you have a WiFi adapter in place you can use the Internet with no need for an Ethernet connection. The relevant software should be loaded automatically on booting so that there's no need for a keyboard or a mouse to be connected. This can free a USB socket on a model 'B' which can be used for a number of purposes, including providing disk storage (see the section about adding USB drives for more details).

In fact, if you use your Raspberry Pi remotely using SSH, VNC or Windows Remote Desktop (each of which have their own sections elsewhere) you can dispense with the need for a display as well, making your Raspberry Pi almost completely wire-free. To remove the final wire, the power supply, see the section about operation on batteries.

Having an Internet connection will allow you to supply your own network services including web servers and disk servers to the rest of your house, or you could program servers of your own using TCP/IP sockets (also described elsewhere).

Raspberry Pi using a WiFi adapter.

Using Bluetooth Devices

Named after an ancient Scandinavian king, Bluetooth is a low-power radio network intended for short-range connection to simple computer peripherals such as mice and keyboards. It's ubiquitous in mobile phones, where the capability to carry audio over Bluetooth is much used, and is also available in many laptops. It can be added to other computers most simply by using a small USB Bluetooth adapter. These can cost as little as £5.

Bluetooth adapters can now be really quite small.

Adding a Bluetooth adapter can be very efficient in terms of making best use of a limited number of USB ports, because it can be used to support both a mouse and a keyboard (and still enable you to exchange files with your mobile phone).

Adapters vary in the strength of the wireless signal they produce and therefore the distance away from them that a Bluetooth device can be. A class 1 adapter will provide in-room connections for about 100m, a class 2 adapter will reach about 10m (probably anywhere in the same room), and a class 3 adapter needs its devices to be no further than 5m away.

We've already discussed the different abilities of the Model A and Model B Raspberry Pi with respect to the amount of power they can provide to USB devices.

Although a number of Bluetooth adapters are supported under Linux and Raspberry Pi not all of them are. If you're thinking of buying a new one you should refer to the list of working devices held at

    http://elinux.org/RPi_VerifiedPeripherals

How they work

Every Bluetooth device wants to pair up with another Bluetooth device for some function or another. All Bluetooth relationships are like this: they involve just two parties who become friends. Initially both devices are suspicious of each other and will demand some kind of authentication when they meet: typically this is in the form of one device inventing a four-digit number which the other device has to prove its user knows (aiming to prove that the owner of both devices knows about the proposed relationship). This authentication phase can be omitted once a device has been told to trust the other party.

There are a fixed set of functions that a Bluetooth device can either make use of or provide to other Bluetooth devices and each is referred to as a Bluetooth 'profile'. In order to use a Bluetooth relationship for some particular function (such as transferring files, for example) the relevant profile must be supported by both devices. The use that one profile makes of a device isn't always compatible with another, so when a device supports a number of different profiles the user has to indicate exactly which use is intended.

This means that making use of a Bluetooth device from Raspberry Pi for some function or another will involve the following stages:

- Turning Raspberry Pi into a Bluetooth device by installing a Bluetooth adapter and associated software.
- Discovering the device.
- Authenticating the device.
- (For convenience) telling Raspberry Pi to trust the new device, so that authentication isn't necessary on every use.
- Selecting (one of) the profiles the device supports (among those also provided for on Raspberry Pi).

Preparation

Raspberry Pi can use Bluetooth devices, although the standard distribution doesn't come with the software required already installed. To install this software and provide a USB Bluetooth adapter, use

    sudo apt-get install bluetooth blueman

If you don't intend to use the Bluetooth manager from the desktop you could omit 'blueman', although below we assume you haven't done this. Note that this will install quite a lot of software (perhaps about 100MB). If you have a 2GB SD card (ie a fairly small one) this may be an issue.

Once you've done this you should also be able to see that a new 'Bluetooth' service has been added to the machine and is now running:

    pi@raspberrypi ~ $ service bluetooth status
    [ok] bluetooth is running.

If so you should now plug the Bluetooth adapter into a connected USB port (this could be via a USB powered hub if you anticipate power problems).

The Bluetooth adapter in place and operational.

The new device should be visible in the list of USB devices that Raspberry Pi can see:

```
pi@raspberrypi ~ $ lsusb
Bus 001 Device 001: ID 1d6b:0002 Linux
 Foundation 2.0 root hub
Bus 001 Device 002: ID 0424:9512 Standard
 Microsystems Corp.
Bus 001 Device 003: ID 0424:ec00 Standard
 Microsystems Corp.
Bus 001 Device 008: ID 0a12:0001 Cambridge
 Silicon Radio, Ltd Bluetooth Dongle (HCI mode)
Bus 001 Device 007: ID 04f2:0963 Chicony
 Electronics Co., Ltd
```

(It's the 'Cambridge Silicon Radio' device in the above.)

GUI setup

Now that the adapter's installed you can introduce Raspberry Pi to a Bluetooth device. In this example we'll add a keyboard (this one is really for an iPad, but works well enough with Linux and cost only a little more than £10).

A Bluetooth keyboard.

When you are in the desktop graphical interface (run the command 'startx' if you are not) you should now be able to see a (blue) bluetooth icon at the bottom right-hand side of the screen:

You should click on this icon to begin

the process. A window will appear. On the right-hand side at the bottom of the window (on the status bar) you should see two icons which flicker as data is sent to and received from your adapter. Just to the left there should be information about how much data is travelling in each direction.

Searching for discoverable Bluetooth devices nearby.

Bluetooth devices use extra power when making themselves known to other devices so it's usual to have to request them to become discoverable for a short while (usually a minute or so). Other devices (such as mobile phones) don't make themselves discoverable either, for reasons of privacy. In the case of the keyboard it's made discoverable by pressing a small button on its base called 'connect'. While the keyboard's in this state the 'Search' button should be pressed at the top left of the window.

After a short while your device should appear in the window. This will neither be authenticated, trusted nor connected. To start the process of correcting this situation click on the device and then click on the 'Setup...' button.

The first thing that'll occur is 'pairing', which is where the device is authenticated by Raspberry Pi so that its existence can be remembered for later. The 'Pairing' window that appears will ask for the best way to generate a four-digit number to challenge the device with for authentication. Some Bluetooth devices have no proper means of giving a user the ability to determine a response and so will require a fixed challenge ('Custom Passkey') to be entered (you expect this to be indicated in the documentation that came with your device). Otherwise simply click on the 'Forward' button to be given a number to enter into the device.

In the case of the keyboard the number is typed on the keys followed by the Enter key.

After a screen informing you that pairing is in progress, hopefully (assuming

Ways to generate an authentication challenge.

authentication succeeds) a connection to the device will be established. Raspberry Pi will determine which Bluetooth profiles it has in common with the device and present you with a choice of how it can be used. For the keyboard it identifies an 'Input Service' to connect it to.

Pressing the 'Forward' button should result in a message indicating that the device is now added and connected successfully.

You may notice, at this stage, that there's a message icon from LXDE that appears in the bottom right-hand corner of the screen. Clicking on it may reveal a message that asks you to authorize the use you've requested for your new Bluetooth device. If so simply click on 'Always accept'.

Selecting the Bluetooth profile to use.

Authorizing the use of the remote keyboard.

Browsing for the file to send.

If you want to avoid having to perform a challenge and response in the future click on the device and then on the yellow star to mark that you trust this device.

Sending files

Depending on the device it may be possible to transfer files to and from it using Bluetooth.

After having introduced the new device as above you can transfer a file by right-clicking on the Bluetooth icon at the right-hand end of the LXDE status bar.

Preparing to send files to a device.

This will allow you to select the device you wish to send files to and then browse for the file you wish to send by clicking on the 'OK' button.

Following that it's likely that you then have a short amount of time to verify that you're prepared to accept the file on the destination device, otherwise the transfer will be abandoned.

Transmitting a file over Bluetooth.

Transfer speed over Bluetooth is not as fast as you may be used to over WiFi or Ethernet.

Fetching files

It's also possible to fetch files over Bluetooth, although this is a little more complex because you may need additional software both on Raspberry Pi and on the Bluetooth device.

Install Bluetooth FTP support

The software to install is responsible for allowing the file manager (called 'pcmanfm') to open files whose name begins 'obex:'. It can be installed using

```
sudo apt-get install gvfs gvfs-fuse gvfs-bin
```

This will give Raspberry Pi the ability to see files provided by the Bluetooth 'OBEX FTP' profile (which allows a Bluetooth device's file system to be inspected). Unfortunately even smartphones don't always support this profile, so you may have to install software on your device to support it. On Android devices an application called 'Bluetooth File Transfer' is one possibility.

Set the Bluetooth browse program

It's likely that you'll need to change the command that 'blueman' (the program responsible for the Bluetooth icon) uses to browse files. By default it uses a program called 'Nautilus', which is a little large for Raspberry Pi. The Raspberry Pi comes with a similar program called 'pcmanfm'. To do this, right-click on the Bluetooth icon again but select 'Local Services....'

Select the 'Transfer' set of options and click on the triangle to the left of 'Advanced' to reveal advanced options. In the box it reveals type in

Running local services.

```
pcmanfm obex://[%d]
```

to replace the default browser.

Updating the browser used for Bluetooth Obex: files.

Browsing

Before you begin ensure that the device you're browsing has the right support for the Obex FTP profile and that it's running any required software.

Then it's possible to browse a remote Bluetooth device using the menu provided by the Bluetooth icon on the status line.

After a bit of time the file manager should open with the files visible on the remote device:

These directories can be navigated exactly as if they were local files and files can be copied backwards and forwards in the normal way. For example, this remote device has a camera and this transfer is copying a picture on to Raspberry Pi:

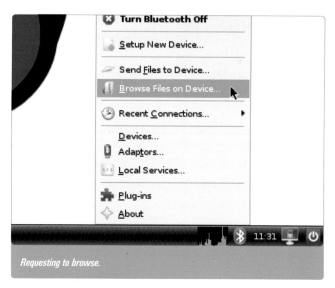

Requesting to browse.

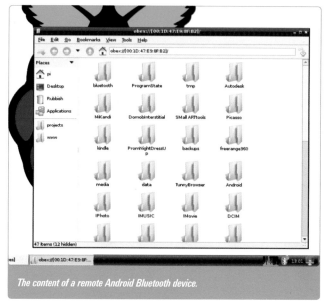

The content of a remote Android Bluetooth device.

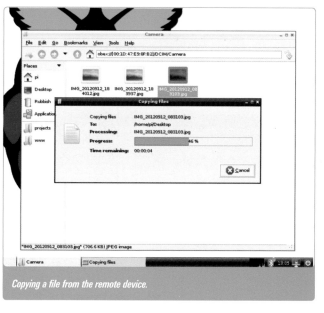

Copying a file from the remote device.

HARDWARE RECIPES
LEDs

Your Raspberry Pi has five Light Emitting Diodes. Three are directly connected to the chip that provides the Ethernet connection and are used to indicate its state. The fourth is directly connected to the power supply and so simply indicates whether power is available. The final LED can be controlled by software.

LED functions

LEDs on Raspberry Pi (revision 1 board).

The function of the different LEDs is as follows:

LED software control

Only one of the LEDs can be controlled by software. The normal Raspberry Pi kernel sets up the LED to reflect activity on the SD card, but this can be changed.

In Linux LEDs are controlled via files located in the /sys directory which is supported entirely by kernel drivers (it doesn't use the SD card or other storage medium). Each kernel driver and device have their own subdirectories (indexed in several different ways) under /sys. One place you can find the directory corresponding to the LED device that's under software control is

```
/sys/class/leds/led0
```

Inside this directory there are various other files and directories that can be used to alter its behaviour or report its status:

```
pi@raspberrypi ~ $ ls /sys/class/leds/led0
brightness   device   max_brightness   power
   subsystem   trigger   uevent
```

LED triggering
Depending on the kernel there may be a number of internal events that an LED can be assigned to monitor. The list of such uses is available by looking at the contents of the trigger file:

```
pi@raspberrypi ~ $ cat /sys/class/leds/led0/
 trigger
none [mmc0]
```

This indicates that there are two possible trigger settings: 'none' and 'mmc0'. The fact that 'mmc0' appears in square brackets means that 'mmc0' is currently selected. This means

Label	Colour	Function
ACT or OK	Green	(Status OK) This LED is controlled by the Raspberry Pi software via General Purpose Input/Output (GPIO) line 16.
PWR	Red	(Power) This LED lights when power is plugged into the board (via the USB power connector).
FDX	Green	(Full Duplex) An Ethernet connection either allows a transmission and a reception to occur over the connecting wires at the same time (full duplex) or requires the wires to be used first for transmission and then for reception without both occurring at the same time (half duplex). This LED will light when full duplex is being used.
LNK	Green	(Link Active) This LED lights whenever data is being transmitted or received on Ethernet.
10M or 100	Yellow	(100Mbps) Data can pass over the Ethernet connection either at 10 million bits per second or at 100 million bits per second. This LED lights if the faster (100Mbps) connection is being used. (The label '10M' was a mistake on the revision 1 Raspberry Pi boards.)

that the LED will flash whenever the MultiMedia Card device number 0 is used (this is the SD card).

To put the LED under software control the trigger source must be 'none'. This is achieved simply by writing 'none' into the 'trigger' file. Because only the privileged 'root' user can write to this file the 'sudo' command will have to be used. For example:

```
echo none | sudo tee /sys/class/leds/led0/
  trigger
```

It's possible to add extra triggers to Linux by loading additional kernel modules. For example, this will load a 'heartbeat' trigger that causes the LED to flash regularly to indicate that the operating system is still running:

```
sudo modprobe ledtrig_heartbeat
```

Following this another trigger becomes available:

```
pi@raspberrypi ~ $ cat /sys/class/leds/led0/
  trigger
none [mmc0] heartbeat
```

which can be used in the normal way.

Turning LEDs on and off

One of the other files in the 'led0' directory is called 'brightness' and is used to alter the brightness of the LED using an eight-bit number where 0 is no brightness at all (off) and 255 is as much brightness as possible (on).

The LEDs on Raspberry Pi distinguish only two brightness values. All non-zero brightness values turn the LED on fully.

To turn the LED on use

```
echo 255 | sudo tee /sys/class/leds/led0/
  brightness > /dev/null
```

and to turn the LED off use

```
echo 0 | sudo tee /sys/class/leds/led0/
  brightness > /dev/null
```

Security

It can be inconvenient (especially in scripts) to use 'sudo' each time the LED must be altered. In principle the user might be asked to supply a password at any time 'sudo' is used (although not with the normal Raspberry Pi setup).

To allow normal users to write to the 'brightness' and 'trigger' files use the command:

```
sudo chmod a+w /sys/class/leds/led0/
  {brightness,trigger}
```

Unfortunately because the /sys file system is recreated every time Linux boots this will only be effective until the next reboot.

One solution to this problem is to place the command above in the file found at

```
/etc/rc.local
```

This file is executed at the end of the boot process as the user 'root'. This can be achieved using

```
sudo editor /etc/rc.local
```

(We discuss how to select your favourite editor in the section about running programs regularly.)

Use from Bash

Assuming that you've overcome the security problem, here's a short example Bash function that will flash the LED every second forever (which you could adapt to do something more useful):

```
flash() {
  local bright=0
  while true; do
    echo $bright > /sys/class/leds/led0/
      brightness
    bright=$((bright ^ 0xff))
    sleep 1
  done
}
```

After you've typed this in type the command 'flash'; type Ctrl-C to stop.

If you are familiar with Bash scripts you may be able to see how this works. Perhaps the least obvious part, though, is '$((bright ^ 0xff))'. Text inside the dollar double brackets (ie 'bright ^ 0xff') is evaluated as an arithmetic expression and the whole thing is replaced by the characters of the resulting number (for example, $((2+2)) would simply be replaced by the single character '4'). The expression here takes the value currently held in the variable 'bright' and the hexadecimal number 'ff' (255 in decimal), and combines them using the exclusive-or function (denoted '^'). The exclusive-or function looks at each argument as if it were a series of bits and returns a value where each bit in the result is 0 if the corresponding bits in the arguments were the same, and 1 if they were different. The result is that when 'bright' contains the eight bits 00000000 (decimal 0) the result will be 11111111 (decimal 255), and when it contains 11111111 (decimal 255) the result will be 00000000 (decimal 0).

So, each time around the loop the value in 'bright' toggles between 0 and 255.

You may wish to incorporate these definitions in your own Bash script to turn the LED on and off:

```
dir_led=/sys/class/leds/led0
original_use=

# take control of the LED
led_claim() {
    original_use=$(sed 's/.*\[\(.*\)\].*/\1/'
      $dir_led/trigger)
    echo none > $dir_led/trigger
}

# relinquish control of the LED
led_release() {
    echo $original_use > $dir_led/trigger
}
```

```
# turn the LED on
led_on() {
    echo 255 > $dir_led/brightness
}

# turn the LED off
led_off() {
    echo 0 > $dir_led/brightness
}
```

The 'sed' command in the above expands to whatever word is enclosed in square brackets in the 'trigger' file.

Use from Python

Again, assuming that the security problem has been overcome you could use the following in a Python program to provide functions to turn the LED on and off:

```python
import re

dir_led="/sys/class/leds/led0"
original_use=None

def led_claim():
    "Take control of the LED"
    global original_use
    trigger_file = dir_led+"/trigger"
    with open(trigger_file, 'r') as f:
        m = re.match('\[(.*)\]', f.read())
    if m:
        original_use = m.group(1)
    else:
        original_use = None
    with open(trigger_file, 'w') as f:
        f.write('none')


def led_release():
    "relinquish control of the LED"
    global original_use
    if original_use:
        with open(dir_led+"/trigger", 'w') as f:
            f.write(original_use)

def led_on():
    "turn the LED on"
    with open(dir_led+"/brightness", 'w') as f:
        f.write('255')

def led_off():
    "turn the LED off"
    with open(dir_led+"/brightness", 'w') as f:
        f.write('0')
```

This could be tested with a main function such as:

```python
import time

def main():
```

```python
    led_claim()
    for i in range(100):
        print("on")
        led_on()
        time.sleep(1)
        print("off")
        led_off()
        time.sleep(1)
    led_release()
    return 0
```

which will turn the LED on and off waiting a second in each state one hundred times.

HARDWARE RECIPES
Connecting to Hardware using GPIO

Your Raspberry Pi can interact with your own electronics by using pins on its expansion header. These pins are general purpose insofar as Raspberry Pi can control how they're used. It can determine which pins are used for input (where the processor can determine whether there's a voltage on a pin), or output (where the processor can decide whether a voltage should be set on the pin). Many embedded systems employ a similar scheme, which is usually referred to as General Purpose Input and Output (GPIO).

The expansion header

This shows where the expansion header can be found on Raspberry Pi. Each of its pins are numbered.

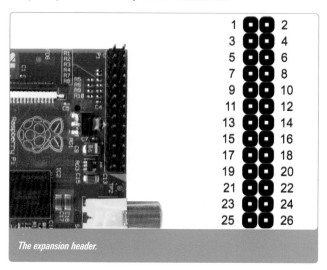

The expansion header.

Not all of the pins on the expansion header are available for GPIO and others may need special configuration before they can be used. With the right configuration the following diagram shows the GPIO lines that could be made available:

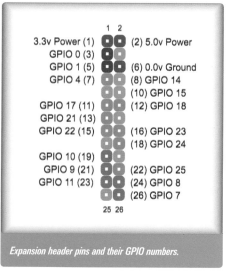

Expansion header pins and their GPIO numbers.

Pins 3 and 5

The two pins marked in dark blue differ in two editions of Raspberry Pi. The diagram shows the correct pins for version two. In the first version these two pins gave access to GPIO 0 and 1 instead of 2 and 3. In both cases, though, these two pins also differ from other pins insofar as they're expected to be used for input, having an 1800Ω pull-up resistor that'll make them read a high value when nothing's connected to them.

Third-party expansion

In the newer version of Raspberry Pi there's also a second set of four GPIO lines that are available from a socket (marked 'P5') optionally mounted underneath the board. These are intended for use by third-party hardware manufacturers.

The second GPIO port on the underside of the PCB.

Looking from the top (which makes the pin number appear backwards) the default use of these pins is as follows:

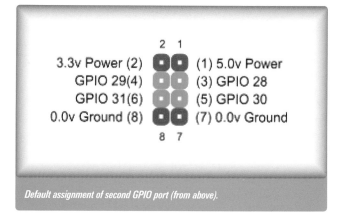

Default assignment of second GPIO port (from above).

Power

Pins 1, 2 and 6 on the expansion header provide different voltages for your electronics (pin 1, marked on the PCB as 'P1', is in the first column on the left on the bottom row).

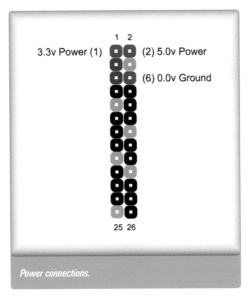

Power connections.

Electrical precautions

Power pins

There's a limit to the current that you can draw from the power pins. No more than 50mA can be taken from the 3.3V pin, but the 5.0V pin can provide whatever current the power supply generates that isn't required by the board. In the Model A board the board takes about 500mA, and the Model B board takes about 700mA. Since most USB power supplies generate 1000mA there may be 500mA available in the Model A and 300mA in the Model B.

GPIO pins

 Warning

Stick to the current and voltage limits for GPIO pins, otherwise you may damage Raspberry Pi.

When a GPIO line is being driven by Raspberry Pi its voltage is set to 3.3V. The amount of current that can be supplied is programmable from 2mA to 16mA (the annexe about configuration gives information about how this is achieved). Normally your hardware shouldn't draw more than 2mA. There's no over-current protection.

When a GPIO line's being used as an input it mustn't be subject to voltages greater than 3.3V. There's no over-voltage protection.

Third-party GPIO boards

In effect the above means that making a mistake in your hardware might destroy all or a part of your Raspberry Pi. There are three main ways to approach this:

- Rely on luck and expertise to stay within these limits.
- Use a third-party interface board with its own 'buffers' (to drive greater currents) and 'level converters' (to deal

with a greater range of input voltages) to handle external electronics.
- Provide buffers and level converters of your own design.

Two popular third-party interface boards are the 'Gertboard', which can be purchased, for example from Element14 at

```
http://www.element14.com/community/groups/
    raspberry-pi
```

and Pi-Face, which can be obtained from the School of Computer Sciences at Manchester University at

```
http://shop.openlx.org.uk/
```

Furthermore there are many others listed here:

```
http://elinux.org/RPi_Expansion_Boards
```

Note, however, that some boards that provide GPIO pins don't use the ones provided directly on Raspberry Pi. Pi-Face, for example, provides GPIO lines but accesses them via Raspberry Pi's SPI interface. The section about building a twitter alert device demonstrates the use of a Pi-Face board.

How GPIO software works

The most basic thing that can be done with a GPIO pin is to determine whether it's to be used as an input or an output. If it's an input, find out whether the electrical signal it's attached to is active or inactive; and if it's an output, set its voltage to be either active or inactive.

Sometimes it's also possible to say which signal levels (out of high and low) you want to consider 'active'. Signals could be considered either active when they're high or active when they're low.

More advanced operations allow the state of an input pin to be monitored in hardware interrupting the computer when it changes. You might be interested in the signal visible at an input pin at two times: when the input on the pin goes active having been inactive ('rises'); and when the input on the pin becomes inactive having been active ('falls'). If you draw a graph of the voltage being monitored, in time such transitions will look like the edges of successive square waves. Consequently these transitions are called 'edges'.

Commonly software will therefore provide ways to:

- Enable and disable a GPIO pin.
- Set which signal level is 'active'.
- Determine whether it's an input or output.
- Write a value to a pin.
- Read a value to a pin.
- Set the edges that will generate input interrupts.
- Wait for an interrupt to occur.

Performance concerns

There are many ways to use signals and in many applications performance isn't really an issue.

One potential concern, however, is the rate, or frequency, at which the signal changes. In the best of possible worlds the highest frequency a pin could operate at would be determined by electrical features (the characteristics of the GPIO circuitry and either the load that it's driving or the resonance in what it's sensing). However, it's quite likely that it's the intervening software that'll actually provide a limit. The time it takes for the software to operate will limit the number of times per second that a change can be requested or that a change can be detected.

For example, if your application uses GPIO to deliver a number of pulses in a certain amount of time (perhaps to change the position of a stepping motor) you must be sure that the software can deliver pulses at the required frequency, otherwise there's a risk that some pulses may be omitted.

Another possible performance problem is the variation in the amount of time, or latency, that it can take to detect a change. Again, software is the most likely source of this variation.

For example, if your application uses the amount of time between a signal becoming active and subsequently becoming inactive as data (perhaps as a sample in a pulse-width-modulated audio stream), it's important that variations in latency aren't as large as an amount of time that would make the measured duration mean one data value instead of another.

Linux software support

GPIO is a common facility to have on embedded computers, so Linux has standard ways to use GPIO hardware. In fact there are standard ways to provide all the functions we mentioned above in (at least) three separate kinds of program.

Kernel-level support

Most tightly associated with the GPIO hardware are the Raspberry Pi GPIO drivers in the Linux kernel. These provide a set of C functions that can be used in other parts of the kernel to create drivers for specific GPIO-using hardware. (Typically a custom device might use a few GPIO lines to communicate with Raspberry Pi: the driver here is one for such a custom device.)

Programs at this level have a very low latency when they read and write to a pin. Unfortunately they're limited by a quantity called the system interrupt latency. This is the potential delay that can be caused by interrupt handling programs that service some other element of hardware. In Raspberry Pi this latency can easily be tens of milliseconds due in part to unfriendly programs in the SD card, USB and Ethernet drivers.

If you're interested in writing your own driver (which will provide the best performance in Linux) the documentation held online here is an excellent reference:

```
http://www.kernel.org/doc/Documentation/
gpio.txt
```

Application-level support

Rather more distant from the hardware, Linux provides a way for user applications to be able to control a GPIO pin via some special files in the 'sysfs' directory. These are not 'real' files corresponding to data on disk. Instead, when they're read or written programs in the kernel are called on the user's behalf.

Enabling, disabling, reading values, writing values, setting which signal level is 'active' and determining which edges will cause an interrupt are all accomplished by normal file operations that read from or write to these files. Waiting for an

interrupt is accomplished using the standard 'poll' or 'select' functions that Unix-like operating systems generally provide to inform the user of any exceptional condition on an open file.

Programs at this level have two additional sources of delay and latency to deal with. One is called scheduling latency. This is effectively the amount of time that it might take Linux to notice that an application program can take its turn to use the processor (other programs having had their turn). Another is system load which occurs when other, more important, application programs need to use the processor and so prevent your program from using it. There's some control over this delay because the importance of your program can be altered using the 'nice' command. Typically, though, both of these delays are most likely to occur while waiting for an interrupt.

Script-level support

The great benefit of having an interface that's implemented almost entirely by reading and writing values to files is that it's easy to use from most scripting platforms (including Bash and Python).

In terms of performance this may be only a little worse than relying directly on application-level support, the main issues simply being that scripts, being interpreted, generally run slower than compiled applications. In the case of 'garbage collected' scripting languages, such as Python, the prospect is always present that instruction execution will halt while unused memory is collected. This effectively adds a variable (but potentially fairly long) latency between the execution of consecutive instructions.

The sysfs interface

To demonstrate how programs at both the application and script level work with the 'sysfs' file interface we'll use Bash, which is the shell normally used on Raspberry Pi.

The directory containing the files that need to be used are all in the directory

```
/sys/class/gpio
```

The contents of this directory are writeable only by the privileged user 'root', so for the purposes of this description we'll assume that the command

```
sudo -s
```

has been executed to log on as 'root'. This is what we expect to see in that directory:

```
root@raspberrypi:/home/pi# ls /sys/class/gpio
export gpiochip0 unexport
```

Two 'files' are in this directory that enable and disable a GPIO pin called 'export' and 'unexport': to start using a given GPIO pin number 23, simply write the text '23' to the 'export' file:

```
echo 23 > /sys/class/gpio/export
```

and to stop using it again later write the same value to 'unexport'.

```
echo 23 > /sys/class/gpio/unexport
```

Between the two you might notice that a new subdirectory has appeared for the enabled pin called 'gpio23':

```
root@raspberrypi:/home/pi# ls /sys/class/gpio
export gpio23 gpiochip0 unexport
root@raspberrypi:/home/pi# ls /sys/class/
  gpio/gpio23
active_low direction edge power subsystem
  uevent value
```

It's the files in this directory that are required for the other functions.

To set which signal level is 'active' on this pin either write '0' (for active high) or '1' (for active low) to the file 'active_low'. For example, to set active high:

```
echo 0 > /sys/class/gpio/gpio23/active_low
```

(In fact you'll find that this value is already '0' to start with.)

To determine whether it's an input or output, write 'in' or 'out' to the file 'direction'. For example, if we want to use GPIO pin 23 to read the signal on it we would use:

```
echo in > /sys/class/gpio/gpio23/direction
```

To read the value currently available on the pin read the file 'value':

```
cat /sys/class/gpio/gpio23/value
```

The character '0' will be returned if the signal is inactive and '1' if it's active.

Conversely if we'd set the pin up as an output we'd write to this file (again '0' for inactive and '1' for active). For example, to make the signal active:

```
echo 1 > /sys/class/gpio/gpio23/value
```

We can easily set the edges that'll generate input interrupts by writing either 'none', 'rising', 'falling' or 'both' to the file 'edge'. For example, to set an interrupt that occurs on both rising and falling edges:

```
echo both > /sys/class/gpio/gpio23/edge
```

To wait for an interrupt to occur we need to call the system function 'select' with a file descriptor for the 'value' file. There's no built-in way to do this in Bash, although it's quite easy to keep trying every second like this:

```
while [ $(cat /sys/class/gpio/gpio23/value)
  -eq 0 ] ; do
    sleep 1
done
```

Software bypassing the kernel

Some operations provided by Raspberry Pi's GPIO hardware cannot be provided through Linux's generic interface. For example, Raspberry Pi can be programmed to provide its own internal pull up or pull down resistor to determine the signal that'll be read from a GPIO line when nothing is connected to it. It's also possible to provide timed pulses on a GPIO line for applications that use pulse-width modulation. There are also a variety of alternative functions that can be selected for each pin which it wouldn't make sense to access via a standard GPIO interface.

To use these features direct access is required to the GPIO hardware. The GPIO lines are controlled by 'registers' with fixed physical addresses in the ARM's memory. It's possible to ignore the GPIO driver built into the kernel and access these registers directly.

Physical addresses aren't normally available to application code (we discussed physical and virtual addresses in the section about operating systems), but there's a method that'll allow the privileged 'root' user to set aside a section of an application's memory that'll be 'mapped' on to specific physical addresses.

A good example of an open and very well commented small library of C functions that'll achieve this and provide useful functions to control individual GPIO pins (and related devices) can be found at

```
http://www.open.com.au/mikem/bcm2835/
```

Software for third-party boards

As we mentioned above, not all third-party boards with GPIO capabilities use the BCM 2835's native GPIO peripherals. Direct access to Linux's GPIO drivers can't normally be used to control these and different software might be required for each different third-party board.

Gordon Henderson's 'WiringPi' library, originally intended to mimic a similar library used for the open-hardware 'Arduino' processor, tries to address this problem by providing a set of identical C function calls that are implemented differently depending on the board used. Currently it provides support for both direct (memory mapped) access to the Raspberry Pi's GPIO lines and 'sysfs' access, as well as access to the GPIO lines on 'Gertboard' and 'Pi-Face'. It's available as an online download from

```
https://projects.drogon.net/raspberry-pi/
  wiringpi/
```

Installing WiringPi

The library is available as an online download on your Raspberry Pi using

```
wget http://project-downloads.drogon.net/files/
  wiringPi.tgz
```

This will need to be unpacked using

```
tar -xzf wiringPi.tgz
```

which will create a new directory called WiringPi. This contains a library, written in C, that provides functions for handling GPIO lines attached to Raspberry Pi. To create the library (using the C compiler) use

```
cd wiringPi/wiringPi
make
```

The library and the C headers that describe the functions it makes available can then be installed to /usr/local using

```
sudo make install
```

In addition to being able to be used by your own programs this library is used by a 'gpio' command which can be used in scripts of your own. To build and install this too, continue

```
cd ../gpio
make
sudo make install
```

Python support

A Python module is available that uses the memory mapped style of interface to Raspberry Pi's native GPIO peripherals (and can therefore access its additional functions). It's available using

```
sudo apt-get install python-rpi.gpio
```

There's also a version for use in Python3 which can be installed using

```
sudo apt-get install python3-rpi.gpio
```

Some documentation can be found at

```
http://pypi.python.org/pypi/RPi.GPIO
```

and more can be provided through Python's help system by installing the code and then running the Python commands

```
import RPi.GPIO as GPIO
help(GPIO)
```

Note that because this library uses memory mapping, and this is something that only the privileged 'root' user is allowed to do, Python programs written using this library need to be run using 'sudo'. For example, to run the above 'help' Python should be run:

```
sudo python
```

More GPIO lines

If you have a project that requires more GPIO lines than the 17 that the expansion header provides it's possible to use GPIO lines that are elsewhere on the board.

The most obvious, on version two of Raspberry Pi, are the additional four (GPIO28, 29, 30 and 31) on the third-party connector 'P5'. Four more (GPIO 2, 3, 5 and 27 on version one boards, and GPIO 0, 1, 5 and 27 on version two) are present in the camera serial interface (CSI) if that isn't being used. On Model A boards a signal used for the missing Ethernet (GPIO 6) is available on its own pad on the printed circuit board.

Even more lines are available if you want to ignore or electrically disconnect their current use. A good list can be found at:

```
http://elinux.org/RPi_Low-level_peripherals
```

The GPIO pins that are provided on the Raspberry Pi circuit board are by no means the full set of GPIO signals that the BCM 2835 system-on-a-chip can provide. Most of these simply cannot be used. Nonetheless, there's complete documentation of each GPIO line, of the alternate functions they support and of how to control them in the chip's data-sheet, which at the time of writing can be found online at

```
http://www.raspberrypi.org/wp-content/
    uploads/2012/02/BCM2835-ARM-Peripherals.pdf
```

It's possible that a web search might provide a more up-to-date version. The above document has an errata (error correction) that's available at

```
http://www.scribd.com/doc/101830961/
    GPIO-Pads-Control2
```

which explains what's meant by the drive strength of a GPIO pin. Note that in these documents the expansion header is referred to as 'The GPIO Connector (P1)'.

Special function pins

Several of the pins can be used either for GPIO or for various special functions including supporting the UART (cyan in the diagram below), an I2C device (dark blue) and two SPI devices (magenta).

Pins 3 and 5 also vary slightly on older version 1 boards because they are connected to a different I2C device (I2C0). The peripherals these special function provide will be discussed in the following sections.

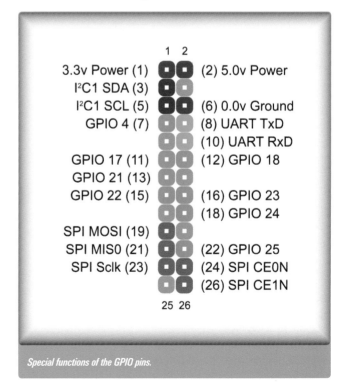

Special functions of the GPIO pins.

Connecting to an SPI Device

Motorola originally defined the Serial Peripheral Interface (SPI) bus which uses four wires to transfer data serially one 'data frame' at a time. Like I²C it's a master/slave protocol that supports multiple slaves, only one of which can talk to the master at any one time. Raspberry Pi fills the master role.

As a protocol, SPI is supported by every SD or MMC memory card, but the more likely use on Raspberry Pi is to communicate with external chips as I²C does. The chips that SPI provides access to, however, are often a little more complex than those using I²C. Furthermore the data transfer rate can be significantly greater. One common use is to provide a connection to a microcontroller board, and it's been used for things like USB controllers and Ethernet adapters. A few of the peripheral boards designed for Raspberry Pi (such as Pi-Face) are accessed via an SPI interface.

Pins
The pins appear on the expansion header as indicated by the magenta colour below:

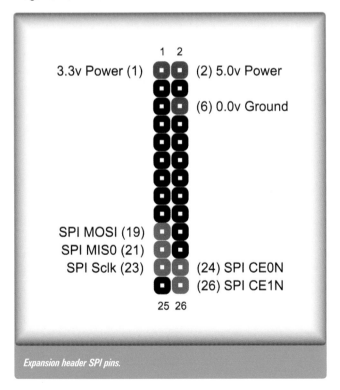

3.3v Power (1)   (2) 5.0v Power

(6) 0.0v Ground

SPI MOSI (19)
SPI MIS0 (21)
SPI Sclk (23)   (24) SPI CE0N
(26) SPI CE1N

Expansion header SPI pins.

How it works
The interface is quite simple. Raspberry Pi (the master) uses one of the pins to set a clock signal (Sclk) which defines when data should be read on the two data lines: one from the master to the slave (Master Out, Slave In: MOSI), and one from the slave to the master (Master In, Slave Out: MISO).

Raspberry Pi supports up to two slaves, only one of which can be active at any one time. One 'Chip Enable' pin is supplied for each of the slaves – the one with no voltage on it identifies the slave that can be active. These two pins are called CE0N and CE1N (CE for Chip Enable and a final 'N' to indicate that the signals are active low). Some devices have their own sub-addressing schemes that effectively allow more devices to use the same chip enable.

Normally data is sent a byte at a time, the bits of a byte being sent one after another, most significant bit first, each time the clock voltage rises. This can happen on both data lines at the same time, so that a byte of data can be sent from the master to the slave at exactly the same time that a byte is sent from the slave to the master.

With some SPI devices it's possible to 'daisychain' more than one SPI device from the same slave, connecting the output data of one slave directly into the input data of the next. In effect Raspberry Pi will see a single device with as many bits available as there are in both real devices.

At the lowest level the SPI 'mode' describes the way that the data and clock are supposed to behave in order to indicate the value of individual bits. There are two aspects of this behaviour that both the transmitter and the receiver must agree on. The first is the polarity of the clock – in particular, when the clock isn't being used is its signal high or low? The second is the clock phase: is the data indicated by the signal level on the data line when the clock changes from low to high (it's rising edge) or from high to low (it's trailing edge)? The combination of the two polarity options and the two phase options makes a total of four possible 'modes'.

Using the Linux driver
The most practical way to make use of an SPI device is to use a Linux driver.

Because the SPI pins on the expansion header are also normally GPIO pins the driver that's available, like the I²C driver, is provided in its own module, called 'spi-bcm2708', and similarly needs to be un-blacklisted to allow Linux to load it automatically. To use the Linux driver your first task will be to edit the blacklisting file using

```
sudo editor /etc/modprobe.d/raspi-blacklist.conf
```

(We discuss how to make 'editor' use your favourite editor in the section about running programs regularly.) Again, your change should place a hash (#) character in front of the driver's name to turn that line into a comment, which will be ignored. The 'spidev' module is also required in order to create entries in /dev for the driver, but that should be loaded automatically.

After rebooting using

```
sudo reboot
```

you should observe that the module has been incorporated into the kernel:

```
pi@raspberrypi ~ $ lsmod
Module             Size  Used by
:
spidev             5136  0
8192cu           485042  0
evdev              8682  2
joydev             9102  0
spi_bcm2708        4401  0
i2c_bcm2708        3681  0
```

('lsmod' lists all the modules installed.)

You should also see that new devices have appeared in the 'sysfs' file system, along with a new directory, '/sys/bus/spi':

```
root@raspberrypi:~# ls /sys/bus/spi/devices/
spi0.0   spi0.1
```

There are two devices because Raspberry Pi has two chip enables for two slave devices. Unlike the I²C device both of these devices should appear in /dev

```
root@raspberrypi:~# ls /dev/spi*
/dev/spidev0.0   /dev/spidev0.1
```

Set the device's access properties

The entries in /dev for the SPI devices can be read and written only by the privileged 'root' user. There are a few things you could do about this:

■ Nothing, just run every program that needs to access the devices using 'sudo'.
■ Change the access controls temporarily using 'chmod +rw /dev/spidev*'.
■ Make the system set the device's access control on every reboot.

We described how to do the last in the section about I²C devices. It requires a rule to be placed in a file in the directory '/etc/udev/rules.d' with its own numeric prefix. For example, you might create a new file called '/etc/udev/rules.d/50-spi.rules' (a Pi-Face utility script uses exactly this file name) and add a rule to it such as

```
KERNEL=="spidev*", GROUP="users",
 MODE="0660"
```

the meaning of which will be interpreted whenever udev finds a new device and creates an entry for it in /dev. For new devices whose name begins with 'spidev' this will set its group number to the one named 'users' and its access controls to those represented by the octal value 660, which

allows read and write access by the file's owner and members of its group, but no access to anyone else.

The effect of this rule will be apparent only once Raspberry Pi has been rebooted.

C programming

The correct way to drive an SPI interface is to open the device representing the SPI bus (eg /dev/spidev0.0), then use the multi-purpose 'ioctl' system call to interrogate and configure the connection and then use special message-passing ioctl calls that send and receive data to and from the device.

```
fd = open("/dev/spidev0.0", O_RDWR);
```

will open the bus device and assign a file descriptor (a small integer) to the variable 'fd'. If 'fd' is negative the open has failed.

The driver has defaults for the SPI mode being used, the number of bits in a word (which will get longer if SPI devices are daisychained) and the SPI mode and the speed of the SPI transfer. These values can either be specified using module parameters at the time when the spidev module is loaded or they can be set explicitly using C functions. The way that kernel module parameters can be set was illustrated when we discussed setting the I2C bus speed in the section about I2C.

The defaults for the SPI mode used when reading and when writing data could in principle be different, so there are two calls that can be made to set the mode. The mode itself can be specified by 'or'-ing together two values for the clock phase and the clock polarity. For the first, SPI_CPHA will be used only if the phase samples the data on a trailing edge instead of a rising one; and for the second SPI_CPOL will be used only if the polarity means that the clock is high when it isn't being used instead of low.

```
int rc1, rc2;
 /* return codes we hope will
 be 0 */
int mode = SPI_CPHA | SPI_CPOL;
 /* clock samples trailing edge, high when
 idle */
int actual_mode;

/* mode used for writing */
rc1 = ioctl(fd, SPI_IOC_WR_MODE, &mode);

/* mode used for reading */
rc2 = ioctl(fd, SPI_IOC_RD_MODE, &actual_mode);
```

The number of bits used when writing can be set. The number of bits provided when reading is determined by the interface though, and so can be retrieved rather than set. It's done using:

```
int rc1, rc2;
 /* return codes we hope will
 be 0 */
int tx_bits = 8;
```

```
/* requested bits per
write */
int rx_bits = 0;
/* bits actually used
per read */

/* request the number of bits to use when
writing */
rc1 = ioctl(fd, SPI_IOC_WR_BITS_PER_WORD,
            &tx_bits);

/* discover the number of bits that are being
used when reading */
rc2 = ioctl(fd, SPI_IOC_RD_BITS_PER_WORD,
            &rx_bits);
```

Similar calls are available to assign (on write) or return (on read) the bit order (most significant or least significant bit first) using the ioctl numbers:

```
SPI_IOC_RD_LSB_FIRST
SPI_IOC_WR_LSB_FIRST
```

Like the other devices,

```
char byte = 0;
n = write(fd, &byte, 1);
```

will write the single byte whose value is in 'byte' to the slave device in one transaction. More bytes could be sent using

```
char bytes[4];
… /* fill in byte[0] etc */
n = write(fd, &bytes[0], 4);
```

The value written to 'n' is the number of bytes actually sent, and if it isn't equal to the number set as the last parameter the call has failed.

```
char bytes[32];
n = read(fd, &bytes[0], 32);
```

will read data that's come from the slave in a single transaction (the size of the buffer 'bytes' must be large enough to hold all the data that's returned). The value written to 'n' is the number of bytes received. If it's negative the call has failed.

Beware that these read and write calls may not do exactly what you expect. One consequence of the way that SPI works is that for every bit that's transferred another has to be received. The 'other half' of the data used in 'read' and 'write' is simply discarded. For example, when using 'write' to send a byte SPI will also have had to send a byte back to Raspberry Pi, but that byte will be ignored.

To make best use of an SPI protocol it's necessary to think about what will be received when something is transmitted and vice versa. To implement this SPI supports a method of transmitting and receiving data at the same time. For example, to send the four bytes indicated by the characters 'HiPi' we need to receive four bytes too. This might be

accomplished using the SPI transferring ioctl 'SPI_IOC_MESSAGE' which needs an argument with type C type 'struct spi_ioc_transfer'.

```
int rc; /* return code we hope will be 0 */

uint8_t send[4] = { 'H', 'i', 'P', 'i' };
uint8_t recv[4];

struct spi_ioc_transfer tr = {
    .tx_buf = (unsigned long)&send[0],
    .rx_buf = (unsigned long)&recv[0],
    .len = 4,
}

rc = ioctl(fd, SPI_IOC_MESSAGE(1), &tr);
```

The constants and data types needed to use this more complicated interface are defined in the header:

```
#include <linux/spi/spidev.h>
```

Setting the SPI bus speed
The annexe about configuration describes how the core clock frequency can be altered. Normally the core clock is set to 250MHz and the SPI clock is set by dividing this by a power of two between zero and 16, yielding possible speeds between 3.8kHz and 250Mhz.

This means that not all speeds requested with

```
int rc1, rc2;
/* return codes we hope will be
0 */
int req_speed = ...;
/* requested speed in
Hz */
int actual_speed = 0;
/* speed actually
provided */

rc1 = ioctl(fd, SPI_IOC_WR_MAX_SPEED_HZ,
            &req_speed);
```

can be set. The speed actually set can be retrieved using

```
rc2 = ioctl(fd, SPI_IOC_RD_MAX_SPEED_HZ,
            &req_speed);
```

Further information
There's good documentation about the use of SPI bus devices provided online, including their use from C application programs, at

```
http://www.kernel.org/doc/Documentation/
 spi/spidev
```

HARDWARE RECIPES
Connecting to an I²C Device

When integrated circuits (chips) become so complex that they need a separate mechanism just to tell the rest of their system what they are and what they can do, hardware designers find I²C – the 'two wire interface' originally devised by Phillips – an attractive way to provide it. In this circumstance, and many others, the number of IC pins that are used is far more important than how fast the data it provides access to can be obtained.

With its two wires an I²C bus can be used to attach a large number of different ICs, so it's become the basis of the Systems Management Bus (SMB) standard. However, to a Raspberry Pi enthusiast its ability to manage all the chips in a computer system isn't really as interesting as the range of functions that have taken advantage of the interface's simplicity to create really cheap sensors and other gadgets.

I²C (a contraction of IIC, which stands for Inter-Integrated Circuit) devices costing a few pence to a few pounds are available which can tell the time, find temperature and light levels, and even convert analogue signals to digital ones that Raspberry Pi can read, and vice versa. Many touchscreens that sit transparently over the display are driven through an I²C interface. Some 'intelligent' Hi-Fi systems provide I²C interfaces to turn components on and off or change the volume at the speakers.

Pins

Raspberry Pi provides two I²C interfaces that can be used. One is on the expansion header and the other is on the camera serial interface (which is less convenient to use, and is really dedicated to a different purpose). Curiously, though, the two interfaces are swapped in the older, version one, Raspberry Pi hardware. The pins for version two Raspberry Pis appear on the expansion header as indicated by the dark blue colour below:

The same pins are used but with the 'other' I²C device on the older revision 1 hardware. (To tell the difference look for some mounting holes: if you have some you have revision 2.)

The pins that are used have an alternative function as GPIO lines, so it's important that they're assigned to their I²C function before they're used. The standard Raspberry Pi distribution initially configures these two pins as GPIO lines.

This particular pair of pins are attached to a positive voltage through a 'pull-up' resistor, which means that the signal on them will always be high unless something on the bus does something to pull it low.

How it works

The two wires in I²C's 'two wire interface' are used for data and a clock (which tells observers when the data line is indicating a bit of data). The rate at which bits can be sent over the interface depends on the rate at which the clock changes,

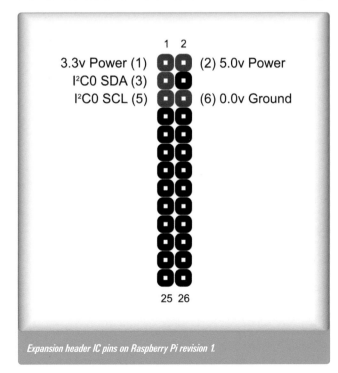

```
            1  2
3.3v Power (1) ■■   (2) 5.0v Power
I²C1 SDA (3) ■■
I²C1 SCL (5) ■■   (6) 0.0v Ground
            ■■
            ■■
            ■■
            ■■
            ■■
            ■■
            ■■
            ■■
            ■■
            ■■
            ■■
           25 26
```

Expansion header IC Pins on Raspberry Pi revision 2.

```
            1  2
3.3v Power (1) ■■   (2) 5.0v Power
I²C0 SDA (3) ■■
I²C0 SCL (5) ■■   (6) 0.0v Ground
            ■■
            ■■
            ■■
            ■■
            ■■
            ■■
            ■■
            ■■
            ■■
            ■■
            ■■
           25 26
```

Expansion header IC pins on Raspberry Pi revision 1.

typically a few tens or hundreds of thousands of times per second (kHz). (In principle the relevant I²C standards guarantee that devices should be able to operate at 100kHz.)

There can be several chips on an I²C bus at once, but only one can be 'in charge' at any one time. This 'master' (Raspberry Pi in our case) uses the bus to send out a signal containing the address of another ('slave') device and an indication as to which direction data is expected to flow between them. Assuming there's a slave with that address it will acknowledge the request before transferring any data, so that the master always knows whether a given address is valid. The data will then flow in the requested direction (either towards the master or towards the slave).

Typically the information that's exchanged between master and slave will request the reading or writing of a small number of values contained in registers inside the slave. In these cases the function of the slave device is likely to be expressed entirely in terms of what happens when its registers are written to or read from.

Software for I²C

As with GPIO devices, there are two main ways that an application could use Raspberry Pi's I²C devices: one is to use a standard Linux device (inside the kernel); the other is to 'memory-map' the BCM 2835 registers that control I²C (outside the kernel). Software that uses the latter approach would need to include code that can manipulate the GPIO functions as well, because of the need to assign the correct function to the I²C pins (which would otherwise be normal GPIO pins).

Although a chip-specific memory-mapped register interface might allow access to chip-specific features it has the disadvantage that any existing Linux code for driving particular kinds of I²C or SMB devices probably can't be used (because it won't expect that kind of interface). It's probably best, therefore, to use the Linux driver.

Using the Linux driver

In Linux, parts of the kernel – in particular drivers for different types of device – can be added to and removed from the kernel while the system is running. These fragments of the kernel are called kernel modules, and there's one providing an I²C driver called 'i2c-bcm2708' (although the chip is called the BCM 2835 it's based on a central core numbered 2708).

For a couple of reasons this driver isn't incorporated into the kernel that's booted. One of them is that when it's used it selects the correct function for the pins in the expansion header, thus preventing their use as GPIO pins.

Un-blacklist the module

The module is held with all of the others that can be optionally loaded in the directory /lib/modules. Normally, however, Linux will install modules whenever it finds one that provides a driver for a device that appears to be present. Normally it would load 'i2c-bcm2708' automatically. To prevent this the reference distribution for Raspberry Pi has placed its name in a special file, '/etc/modprobe.d/raspi-blacklist.conf', that will prevent it being used.

If you want to use the Linux driver your first task will be to remove its entry. This can most conveniently be done by editing the file using

```
sudo editor /etc/modprobe.d/raspi-blacklist.
 conf
```

(We discuss how to make 'editor' use your favourite editor in the section about running programs regularly.) Your change should place a hash (#) character in front of the driver's name to turn that line into a comment, which will be ignored.

After rebooting using

```
sudo reboot
```

you should observe that the module has been incorporated into the kernel:

```
pi@raspberrypi ~ $ lsmod
Module                    Size Used by
:
spidev                    5136 0
8192cu                  485042 0
evdev                     8682 2
joydev                    9102 0
spi_bcm2708               4401 0
i2c_bcm2708               3681 0
```

('lsmod' lists all the modules installed.)

You should also see that new devices have appeared in the 'sysfs' file system (which figured so largely in the way we proposed to use GPIO in that section), along with a new directory '/sys/bus/i2c':

```
pi@raspberrypi ~ $ ls /sys/bus/i2c/devices
i2c-0  i2c-1
```

Put the device in /dev

What you might not see, however, is any entry in the standard location for devices '/dev'. A separate module must be loaded in order to provide entries in that directory:

```
sudo modprobe i2c-dev
```

Once that's been incorporated you should see a new directory in 'sysfs' called '/sys/class/i2c-dev' and two new entries in /dev:

```
pi@raspberrypi ~ $ ls -l /dev/i2c*
crw-rw---T 1 root i2c 89, 0 Oct 15 13:51
 /dev/i2c-0
crw-rw---T 1 root i2c 89, 1 Oct 15 13:51
 /dev/i2c-1
```

If you want to use these devices but don't want to type the 'modprobe' command each time you reboot it's possible to put 'i2c-dev' in the list of modules that are always to be loaded on boot-up by editing this file

```
sudo editor /etc/modules
```

and including a line that just has

```
i2c-dev
```

on it.

Set the device's access properties

You may have noticed that the entries in /dev for the I²C devices allow them to be read and written only by the privileged 'root' user. There are a few things you could do about this:

- Nothing, just run every program that needs to access the devices using 'sudo'.
- Change the access controls temporarily using 'chmod +rw /dev/i2c-*'.
- Make the system set the device's access control on every reboot.

The last involves the 'udev' subsystem which is used every time Linux notices a new device in the system. Normally udev will simply create a new entry in '/dev' when this happens, but it's possible to provide it with a rule that specifies other things to do as well. These rules can be specified by placing them in a file in the directory '/etc/udev/rules.d' with its own numeric prefix. For example, you might create a new file called '/etc/udev/rules.d/51-i2c.rules' (a utility script that's provided with Pi-Face uses exactly this file name), and add a rule to it such as

```
KERNEL=="i2cdev*", GROUP="users",
 MODE="0660"
```

Udev rules are written on a single line in a rule file. They consist of a condition followed by a comma-separated list of actions to undertake when the condition is found to be true. Paraphrasing, the line above says 'if you find a new device whose name begins with 'i2dev' (create a device file in /dev for it and) set its group number to the one named 'users' and its access controls to those represented by the octal value 660, which allows read and write access by the file's owner, and members of its group, but no access to anyone else'. If you wish to you could create a new group that contains only those who are allowed to use I²C devices; make sure that you're in it, and then use that group instead of 'users'.

Setting the I²C bus speed

It shouldn't normally be necessary to alter the speed chosen by the i2c_bcm2708 module, but if your application needs a particularly slow I²C bus, for example, it's possible to set a 'baudrate' parameter.

If the module is loaded manually using modprobe it can be set during that command. For example, to set it to 1000 baud (there is no lower limit for this speed) the following could be used:

```
modprobe i2c_bcm2708 baudrate=1000
```

To use that parameter on each occasion it's possible to write a file that will be used at boot time as follows:

```
echo "options i2c_bcm2708 baudrate=1000" |
 sudo tee -a /etc/modprobe.d/i2c_bcm2708
```

Use the right device

Because the pins used for I²C on the expansion header are different between revision 1 and 2 Raspberry Pis you'll find that these pins correspond to a different Linux device.

Revision	Device on header
1	/dev/i2c-0
2	/dev/i2c-1

Command-line tools

Linux provides some standard command-line tools that can be used to inspect and manipulate devices attached to an I²C bus. To install them use

```
sudo apt-get install i2c-tools
```

They include the following commands:

Tool	Purpose
i2cdetect	Find the devices with addresses on a given I²C bus.
i2cget	Read a value from one of the registers in a specific I²C device.
i2cset	Write a value to one of the registers in a specific I²C device.
I2cdump	Display the values currently held in a range of registers in a specific I²C device.

For example, if you have no I²C devices on bus 0 the following list will show the fact:

```
root@raspberrypi:/install/piface/piface/
 scripts# i2cdetect -y 0
     0  1  2  3  4  5  6  7  8  9  a  b  c  d  e  f
00:          -- -- -- -- -- -- -- -- -- -- -- -- --
10: -- -- -- -- -- -- -- -- -- -- -- -- -- -- -- --
20: -- -- -- -- -- -- -- -- -- -- -- -- -- -- -- --
30: -- -- -- -- -- -- -- -- -- -- -- -- -- -- -- --
40: -- -- -- -- -- -- -- -- -- -- -- -- -- -- -- --
50: -- -- -- -- -- -- -- -- -- -- -- -- -- -- -- --
60: -- -- -- -- -- -- -- -- -- -- -- -- -- -- -- --
70: -- -- -- -- -- -- -- --
```

Creating I2C devices using sysfs

The procedure above effectively registers an I²C bus with the kernel; it doesn't create a Linux device to represent anything connected to the bus. Unlike other buses in Linux there's no automatic enumeration of devices on an I²C bus. The objects on it have to be declared explicitly.

To do this the subdirectory of /sys/class/i2c-adapter containing an entry for the relevant bus is used. For example, with a revision 1 Raspberry Pi you would use the i2c-0 subdirectory.

If you've attached a new I²C device to your Raspberry Pi at address 0x50 which contains a real-time clock, and you'd like to refer to the device as 'wallclock', this could be achieved using the command

```
echo wallclock 0x50 | sudo tee /sys/class/
   i2c-adapter/i2c-0/new_device
```

It will create a new device entry at

```
/sys/bus/i2c/devices/i2c-0/0-0050/
```

which will be deleted by the following

```
echo 0x50 | sudo tee /sys/class/i2c-adapter/
   i2c-0/delete_device
```

Programming using C
The simplest way to drive an I²C interface is to open the device representing the I²C bus (eg /dev/i2c-1), use a special 'ioctl' call to set the device address and then just use 'read' and 'write' to send data to and receive data from the device.

```
fd = open("/dev/i2c-1", O_RDWR);
```

will open the bus device and assign a file descriptor (a small integer) to the variable 'fd'. If fd is negative the open has failed.

```
rc = ioctl(fd, I2C_SLAVE, 0x50);
```

will select the I²C device with address 0x50 as the slave to use in subsequent uses of the file descriptor, and pass back a return code to 'rc'. If 'rc' is negative this operation failed (eg there may be no I²C device at that address).

```
char byte = 0;
n = write(fd, &byte, 1);
```

will write the single byte whose value is in 'byte' to the slave device in one transaction. More bytes could be sent using:

```
char bytes[4];
… /* fill in byte[0] etc. */
n = write(fd, &bytes[0], 4);
```

The value written to 'n' is the number of bytes actually sent, and if it isn't equal to the number set as the last parameter the call has failed.

```
char bytes[32];
n = read(fd, &bytes[0], 32);
```

will read data that's come from the slave in a single transaction (the size of the buffer 'bytes' must be large enough to hold all the data that's returned). The value written to 'n' is the number of bytes received. If it's negative the call has failed.

```
close(fd);
```

returns any resources associated with the file descriptor when you've finished with it.

This simple interface will create one I²C transaction per read or write, which is good for a large range of devices, but some require reading and writing to occur in the same transfer. For these devices there's a marginally more complex interface available containing a dozen or more C functions each prefixed with 'i2c_smbus_' defined in the header:

```
#include <linux/i2c-dev.h>
```

which is available with these function definitions only once this package has been installed:

```
sudo apt-get install libi2c-dev
```

You may otherwise find that the header exists but doesn't include the function definitions.
Documentation for it can be found at

```
http://kernel.org/doc/Documentation/i2c/
   smbus-protocol
```

Programming using Python
Python has three modules that help greatly in using the same calls that C uses to drive I²C.

The first is the 'posix' module, which provides Python versions of the functions 'open', 'close', 'read' and 'write' (all with the same names).

The others are the 'fcntl' and the 'struct' modules, where 'fcntl' provides the 'ioctl' and 'struct' allows bytes to be packed into values for 'write' and unpacked from values retrieved from 'read'.

If you don't want to do it yourself there are unofficial Python packages available on the web that package these modules up for you. One is 'Python-i2c', which allows bytes to be written to and read from as if each I²C device on the bus were an array indexed by device addresses. It can be found, inspected and downloaded at

```
https://github.com/HappyFox/Python-i2c
```

Further information
There's good documentation about the use of I²C bus devices provided online, including their use from C application programs, at:

```
https://i2c.wiki.kernel.org/index.php/
   Main_Page
```

A list of examples of I²C use on Raspberry Pi for robotics is being maintained at

```
http://www.robot-electronics.co.uk/htm/
   raspberry_pi_examples.htm
```

HARDWARE RECIPES
Connecting to a PC using the UART

A serial line, such as that supported by Raspberry Pi's UART (which we introduced in the section about serial lines), was once the main way that users would interact with a computer. Users would sit at a terminal device that simply sent text serially (one character after another) to the computer and displayed the text that the computer sent back. Once computers routinely had their own graphics support the WIMP (Windows, Icons, Mouse and Pointer) paradigm took over and the UART fell to another use: driving the modem (short for 'modulator/demodulator') that provided Internet access over a phone line. Nowadays that use is also a thing of the past and the telltale nine-pin 'serial line' socket has disappeared from the back of new PCs.

Reports of its death are, however, greatly exaggerated. A UART only really needs three wires to work … and it needs virtually no software to drive it. This means that all kinds of embedded computer systems provide one, even if it's only intended to be used by the software developers who bring the platform to life. Often computer systems provide UARTs that aren't actually connected to anything once the system's in full production, and Raspberry Pi isn't much of an exception.

You can gain access to the text-based terminal interface of your Raspberry Pi from a serial cable in the same way that SSH or a keyboard and display can be used to access it.

Expansion header pins

You can use three of the pins from the GPIO block to support UART signals. As we described in the section about the GPIO, the UART pins are provided by one of the alternative functions of the GPIO lines that the chip labels 14 and 15.

The signals implemented are just those that transmit serial data from the Raspberry Pi (TxD) and that receive serial data back (RxD).

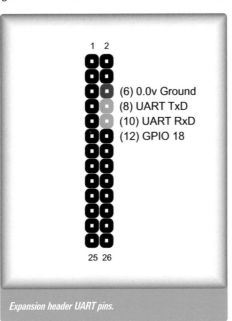

1 2

(6) 0.0v Ground
(8) UART TxD
(10) UART RxD
(12) GPIO 18

25 26

Expansion header UART pins.

A serial connection to a PC

There's no 'proper' UART socket on Raspberry Pi. There's every chance that there's no 'proper' UART socket on your computer either, so this section is going to describe how to use a cheap USB-to-serial cable to connect to your PC using one of its USB (Universal Serial Bus) sockets.

In the section about the GPIO pins you may remember that we have to be fairly careful about the electrical specifications of things we plug in to them. The type of USB-to-serial lead we need is one that supports 3.3 volt connections. The one used in this example has the model number TTL-232R-3V3 and is provided by a company called FTDI Chip.

USB serial lead.

Create a USB serial line

For this project you'll need a TTL-232R-3V3 USB-to-serial lead, a soldering iron, some wire cutters and some solder. The lead has a USB plug at one end and a black 'header' with six single pin sockets in it at the other.

Header end of the lead, wire cutter and soldering iron.

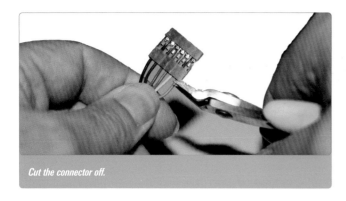
Cut the connector off.

Once cut there should be enough room on both sides of each cut to strip and 'tin' the wires (that is, soak the wires lightly in solder).

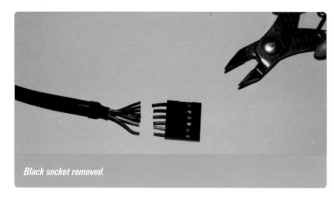
Black socket removed.

Strip the plastic away from the yellow, orange and black wires on the lead and tin them.

Tin the red, yellow and black wires on the cable.

Now also tin the black, brown and red stubs of wire on the header.

Tin the wires on the socket.

It's now necessary to reconnect the wires on the cable to the stubs on the header in the order: (header) black to (cable) orange; (header) brown to (cable) yellow; and, (header) red to (cable) black. We have no use for the other wires. They can be bent out of the way.

You should be able to do this by lining each pair of wires to be connected next to each other (touching) and then applying the soldering iron so as to melt the tinning on each wire.

Reconnected socket.

With a little care you may find it possible to fashion a tiny strip of insulating tape to wrap around the solder join on each connection. Alternatively you might try a small quantity of Sellotape – the purpose of this is to gain some confidence that adjacent connections cannot make electrical contact with each other.

Testing your connection

If you have a Windows PC you can test your connection quite easily because Raspberry Pi sends messages to its UART while it boots.

Connecting to Raspberry Pi

Once you've made up your serial lead it's particularly important that it's connected properly. Like other serial leads we could use, only three wires are connected, even though yours has a six-pin header: the wires actually used in this cable are a black one, a yellow one and an orange one. The header must be connected as shown in this diagram:

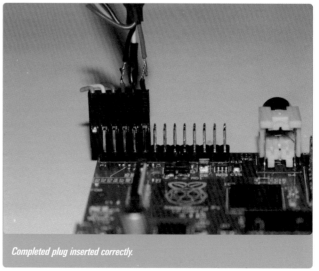
Completed plug inserted correctly.

Calling the pin in the corner of the board 'pin 1' you should be able to see that the header is positioned with its first wire in pin 3 (not pin 1). The black, yellow and orange wires will actually be connected to pins 3, 4 and 5 respectively.

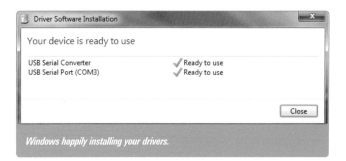

Driver Software Installation

Your device is ready to use

USB Serial Converter   ✓ Ready to use
USB Serial Port (COM3)   ✓ Ready to use

Close

Windows happily installing your drivers.

PC driver installation

Before it can be used on your PC, Windows will need a driver to use the USB-to-serial cable (drivers were described in the section about operating systems). Unless you're using an old version of Windows you should find that Windows will be able to find the driver by itself. Just plug the USB cable into a spare USB socket on your PC.

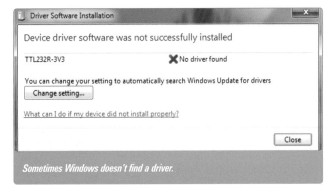

Sometimes Windows doesn't find a driver.

Allow Windows Update to search for a driver on the Internet:

Windows should find a driver on the Internet.

If Windows doesn't find a driver you should be able to find the drivers for our example cable here:

```
http://www.ftdichip.com/Drivers/VCP.htm
```

If you use this address, there's a table of relevant Windows and Linux drivers. You should download and install the driver for your operating system (either 32-bit Windows or 64-bit Windows: to find out which you have look at your System Properties, perhaps by typing on your keyboard the Windows key and the Pause or Break key at the same time).

When a driver has been found you should see a message such as:

In this message you can see the identification of the serial ('com'munications) port (COM3 in the example). You will need to know this information later.

Choosing a terminal emulator

In order to see text being sent to the serial port, or to send any to Raspberry Pi, you need the type of program called a terminal emulator.

If you have an older version of Windows you may find that under Start > Programs > Accessories you'll be able to find the terminal emulator 'Hyperterminal', which you could use. However, since this isn't supported in Windows 7 and later, and since there are better alternatives, we'll describe one of those.

The PuTTY program is free, and has useful features we refer to elsewhere.

PuTTY

PuTTY can be found at this web address:

```
http://www.chiark.greenend.org.uk/~sgtatham/
 putty/download.html
```

Type this into your web browser and download the file 'putty.exe'. Once downloaded run the program to install it (remember that it's good practice always to scan files you've downloaded from the Internet for viruses first).

You should then be able to run the program by following the menu chain Start > Programs > PuTTY > PuTTY.

This program can connect to different computers using a small variety of mechanisms. It remembers details about each program as a 'session'. Before you can use it to access our new serial line you'll need to create a new session.

Start by typing a name for your new session (for example, 'RaspberryPi'), make sure the 'Serial' radio button is selected, and type the speed that Raspberry Pi runs at into the 'Speed' box (115200 baud). Now type the identification of the serial port we found during installation into the 'Serial line' box.

If you've forgotten what the identification of your USB serial port is you can find it as follows. First obtain a Windows command prompt (Start > Programs > Accessories > Command Prompt), then unplug your USB cable and type the command 'mode'. You'll see a list of the serial ports Windows knows about – you can scroll up and down to get the full list.

Now plug the USB cable in again and repeat the process, noticing the name of the new device. (That is the name to use in PuTTY.)

You should now press the 'Save' button on the PuTTY configuration dialogue. These settings will now be available whenever you run PuTTY and double-click on 'RaspberryPi' in the list of saved sessions. Which is what you should do next.

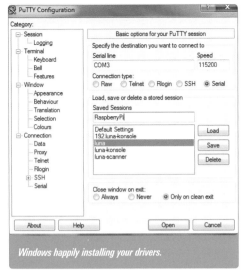

Windows happily installing your drivers.

Use 'mode' in a command prompt window.

The moment of glory
Now plug the power into Raspberry Pi. After a brief pause you should see text flood on to the PuTTY window. Eventually the text will come to an end with an invitation for you to log on over the serial line.

You should type a valid name and password. The reference SD card has the name 'pi' with password 'raspberry' already set up.

HARDWARE RECIPES
Operating using Batteries

Think of all the mobile gadgets you could build if you could free Raspberry Pi from its wires! Robots and radio-controlled models could become autonomous, interactive presentations could become pocket-sized, measurements could be made in strange environments and hardware demonstrations could go mobile.

We've already covered how Raspberry Pi can be communicated with using WiFi or Bluetooth in other sections. All that's really needed is a portable power solution, and taking power from a battery pack is the obvious one. The bad news, however, is that things aren't that simple.

Misinformation
Early proposals for Raspberry Pi included an additional piece of hardware called a voltage regulator, which allowed it to be powered from a very wide range of sources. While producing the board at the right cost lost that capability it didn't revoke some of the early promotional information which said 'Yes. The device should run well off 4 x AA cells.' Sadly this isn't really the case.

Voltage problems
In fact without a voltage regulator Raspberry Pi is really quite sensitive about its power supply. It should have a voltage between 4.75V and 5.25V. Higher voltages aren't guaranteed to be safe. You should consider that this voltage is fed unmodified to the USB boards and to the HDMI connector. Hardware connected to those interfaces might not be able to cope with voltages outside this range either.

The great thing about Raspberry Pi is that it's cheap. If a high voltage destroys it another can be bought with only a modicum of embarrassment. If the living room TV connected to the HDMI port stops working, however, the rest of your family may think differently.

There have, in fact, been many reports of people driving Raspberry Pi at 6V or so and surviving undamaged. Even in these cases, though, the higher voltage is likely to have shortened Raspberry Pi's lifetime.

Batteries of different types go flat at different rates. Some may provide the same voltage for a long time and then suddenly drop dramatically, others may drop little by little over their lifetime. In either case, once the voltage becomes too low Raspberry Pi will start to misbehave in various entertaining ways, the least attractive of which could result in the corruption of the file system stored on the SD card. The drop in voltage can be a particular problem if you try to address the high voltage problem by using partially discharged batteries.

Current problems

To make the picture a little more difficult, although the main chip in Raspberry Pi was designed for life in a mobile phone the whole board wasn't. It expects the power supply to be able to provide a relatively high current, 700mA, and some batteries may find this difficult. In particular there's a property of batteries, called internal resistance, which means that the greater the amount of current they supply the lower the voltage it creates. When Raspberry Pi has a fluctuating workload this may mean that the current, and thus the voltage, goes up and down taking it in and out of specification as it does so.

A relatively high current also makes the cable between the power supply and the Raspberry Pi quite important. If it's a poor quality one its resistance will waste some of the voltage in (imperceptibly) heating the cable up. We've seen cables that drop half a volt – a critical amount when the range of acceptable voltages is itself half a volt.

As we've discussed earlier, many USB devices take their power directly from the USB socket they're plugged into. The current required from this is taken directly from Raspberry Pi's power supply and so will represent an additional drain on batteries and also exaggerate any voltage drop due to internal resistance in the battery or a poor cable.

Energy problems

A common report is that when Raspberry Pi runs on four batteries it'll consume about 375mA from each. You'll find on many batteries a power rating in terms of how much current it can provide for an hour. Rechargeable 'AA'-size nickel metal hydride (NiMH) batteries can be found rated at 1500mAH quite easily, and might give four hours' life.

If this seems low for your gadget you have an energy problem, which you can change into a battery-expense problem by buying the same-size batteries with a larger capacity, or a weight problem by buying bigger batteries.

Unmodified battery packs

Despite all the unfriendly warnings above you may still wish to make up a battery pack and attach a micro USB plug to it to power your Raspberry Pi.

There are several different kinds of battery on the market. The most available are

Type	Cost	Voltage
Zinc-carbon (AA, AAA, etc.)	Cheap	1.5V
Alkaline	Moderate	1.5V
Nickel metal hydride (NiMH) (AA, AAA, etc.)	Expensive (rechargeable)	1.2V
Lithium ion (used in PCs, phones, cameras)	Very expensive (rechargeable)	3.2V–3.6V

The task is to find a combination of batteries (of the same type) that will provide a voltage between 4.75V and 5.25V. Normal non-rechargeable batteries will provide either 4.5V, which is too low, or 6V, which is dangerously high.

Rechargeable NiMH batteries, however, should provide 4.8V, which is just within Raspberry Pi's tolerance. However, it's very close: any drop in voltage may bring the power

supply out of specification, so good quality cables are worthwhile and it's a good idea to recharge the batteries a bit before the end of their natural life.

The good news is that NiMH provide a constant voltage for most of their life, dropping off rapidly only during the last 20% or so. The bad news is that a freshly charged NiMH battery can have a slightly higher voltage of up to 1.4V, which would provide too high a voltage (5.6V). After charging, the batteries' voltage drops quite quickly (days), even if the battery isn't used, so one strategy is to have a supply of already charged batteries that have been on the shelf for a few days.

Battery pack with a regulator

A voltage regulator is a small device that provides a constant voltage for as long as it's provided with a higher voltage (which might even vary a little).

For £2 or £3 you can buy a Low-DropOut (or 'LDO') voltage regulator online that can provide up to 1000mA at 5V. In order to use it you'll have to attach it to a battery pack (perhaps four non-rechargeable batteries or five rechargeable ones) and solder it to your own lead.

Although cheap, the disadvantage to this scheme is in the way a voltage regulator works. The correct voltage is provided by wasting the unwanted voltage as power. This means not only that the voltage regulator can become rather warm but also indicates that a certain amount of the power in your valuable batteries is simply being lost, which will deliver a shorter lifetime than you might expect.

Battery pack with DC-DC converter

If you've dabbled in electronics you may know that a transformer can convert any alternating voltage into any other. For example, the right transformer could reduce 6V down to 5V, but only if the voltage is alternating (which means varying, usually quite quickly). What's more, very little power is lost in the process.

The voltage provided by a battery doesn't vary, but a clever device called a 'switched mode DC to DC converter' solves this problem by chopping it up to make the voltage alternate, running it though something that behaves rather like a transformer, and then converting it back into a constant (but different) voltage.

Although more complex than a voltage regulator these

A DC-DC converter with USB connection.

devices waste less of your battery's power and so will make them last longer.

Because the conversion could either be to a higher or a lower voltage there are two choices of design. You could get an up-voltage DC-DC converter, which would allow you to use only one small (and light) battery; or you could get a down-voltage DC-DC converter, which would allow you to use more batteries (and so provide a greater run time).

DC-DC converters vary quite a lot in price. A down converter from 8V or more (eg from six 1.5V batteries) which can provide 1000mA can cost only a couple of pounds.

With some converters you may have to solder together your own lead. The amount of soldering can be minimized if you buy a converter with a built-in USB socket. Such things can be purchased for about £4 and are most obviously available on eBay.

A do-it-yourself 2500mAh battery pack.

Off-the-shelf converters

Radio-controlled model hobbyists have a similar problem to providing battery power that Raspberry Pi has, and they use what they call a 'universal battery eliminator circuit' (UBEC) or a 'switched battery eliminator circuit' (SBEC). These

12000mAh off-the-shelf battery pack.

Showing Granny your latest photos.

are normally either linear voltage regulators or switched mode DC-DC converters as we've discussed above. However, they have the advantage of ready availability, cheapness and coming ready-assembled with leads. Unfortunately the leads are bare, so some form of attachment, and probably soldering, will still be required.

USB battery packs are available quite widely, normally sold as a portable means to recharge phones. Many contain rechargeable batteries. They tend to be quite large (larger than Raspberry Pi) but will require no soldering. Unfortunately an 8000mAH pack costs about the same price as a Raspberry Pi.

Halting

Raspberry Pi does have a 'low power' mode: it's the one that it enters after you run the command.

```
sudo halt
```

In the section about Linux we said this command will turn Raspberry Pi 'off'. In fact it isn't completely off, merely in a low power mode. In principle Raspberry Pi could be turned on again by one of its internal timers or by an interrupt on a GPIO line. In order to provide these features, however, a new version of the file '/boot/bootcode.bin' would be needed.

When it's in low power mode Raspberry Pi should last about three times as long as it would if it were running.

Because it isn't 100% safe to turn Raspberry Pi off while it's running, another power-relevant use of low power mode is to put it in a state where it's safe to turn the power off completely (which will result in much lower battery use). This would require the addition of your own hardware to control the power provided to Raspberry Pi through its USB connection. Ideally such hardware would respond to a 'turn off in a minute' signal and 'turn on again after this amount of time'. Raspberry Pi could then wake up regularly to perform some task and then put itself to sleep again in a way that would extend battery life extensively. (This would be useful, for example for a data-logging project.)

06.
Meal plans

MEAL PLANS
An MP3 Web Server

As we demonstrated in the section about web servers, it's fairly simple to add your own Python programs that can be accessed using a web browser. Python includes many ready-built libraries that can be drawn together to assemble quite useful web applications. In this section we'll put together a simple web server that streams music to your music-playing device (for example a desktop computer).

Preparation

MP3, which is audio format specified for use in 'layer 3' of the MPEG-2 video standard (from which it gets its name), is the de facto standard in which music files are kept, but it isn't the only one.

Although you may not be aware of it, MP3 is patented and requires a licence to be paid before it can be used. Technically, you are subject to licensing fees if you are playing MP3 files on Raspberry Pi.

One common open format, for which no licence is required, is the 'Vorbis', which is normally produced in container files conforming to another open standard, 'Ogg'. In addition to the standard modules that the reference Raspberry Pi distribution normally contains we can support the playback of 'Ogg Vorbis' files by installing these packages:

```
sudo apt-get install python-ogg python-pyvorbis
```

This will enable you to be good (if you so choose) and play only music encoded in a 'free' format.

Introducing Edna

Edna is a reasonably-sized Python program written by Greg Stein which he has kindly placed under the 'open' GPL licence so that you can download and update it if you want to.

It uses several Python modules to help it create a web server that allows users to browse a collection of music files and to stream them to a player.

You can discover its latest version by visiting

```
http://edna.sourceforge.net/
```

which you can do on your Raspberry Pi.

Unpacking Edna

If it was, say, version 0.60, you should make it a home on your Raspberry Pi using

```
sudo mkdir -p /opt
chgrp users /opt
chmod g+rw /opt
mkdir /opt/edna
cd /opt/edna
```

and then download the most recent version using

```
wget http://downloads.sourceforge.net/edna/
edna-0.6.tar.gz
```

This is a gzipped tar file containing a directory containing the program files. It can be unpacked using

```
tar -zxf edna-0.6.tar.gz
```

which will create a new directory called 'edna-0.6' that you can enter using

```
cd edna-0.6
```

A look at the code

The main program is held in edna.py, although it does use additional code in some of the subdirectories. You could inspect it with 'less', your favourite editor 'editor' (we discussed how to set your favourite editor in the section about executing code regularly), or 'Geany' if you want a graphical display.

Edna uses the standard 'BaseHTTPServer' Python module to perform most of the web server work and uses other modules like 'ConfigParser' to make it easy to deal with a fairly complicated configuration file, which is supplied in 'edna.conf'.

Two Python modules are provided: ezt.py and MP3Info.py. The first of these ('easy templates') parses web page templates that enable information to be provided either as HTML or as XML (which we discussed in the section about web scraping). This is used to produce various statistics. The second decodes an MP3 file and is capable of returning various information about it (in particular the main program uses it to find its size, running time and technical details about how quickly it should be played).

Configuring Edna

Before it can be used there are some details you'll need to establish in the configuration file 'edna.conf'. Open this in your favourite editor, for example

Editing Edna's configuration – port number.

The first item that you might consider changing is the port number that the server will use. Most web servers use port 80, which is why this web server allows you to choose another one (so that it can be run alongside a more traditional web server). If you already have another web server (for example providing insults, as described in the section about Python) that uses port 8080, you may need to change this port (e.g. to 8081 or 8888 etc.).

The main item that you need to set, however, is simply the directory in which you'll store your music files.

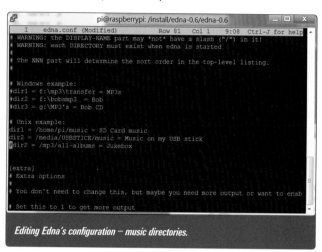

Editing Edna's configuration – music directories.

You can set a number of sources up. The above example change uses only the directory /home/pi/music and another directory that is located on a USB stick.

Once you've finished your changes you should save the amended file.

Running the music server

Making the server run couldn't be simpler. Just execute this command:

```
python edna.py
```

This will start the web server and will also print any activity or problems to the screen. It will keep running forever until the program is closed (perhaps by typing Ctrl-C).

If you want it to run silently in the background you could run it using

```
python edna.py > edna.log 2> edna-errors.log &
```

instead, which will run the web server in the background, keeping a log of its activities in the file edna.log and a list of any errors it encountered in edna-errors.log.

To stop Edna when started like this use

```
kill %1
```

Using Edna

To use Edna you'll need to know the IP address of your Raspberry Pi and remember the port number the service is running on. Edna's web address will be 'http://<IPaddress>:<port>'. In the example below this is

```
http://10.177.67.58:8080
```

Type this into your web browser and you should see a screen like

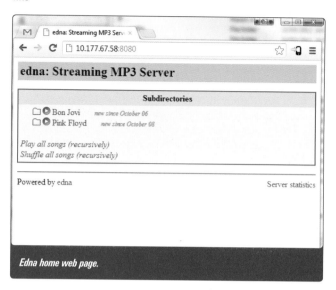

Edna home web page.

It should show you a list of the music files and directories that you have in your music locations. You can choose either to enter each directory and choose among what is found there or you can play all of the songs it contains (with 'Play all songs' or 'Shuffle all songs').

Playing a specific file will result in whatever action your web browser takes when it receives an .m3u (MP3 stream) file. If it uses Windows Media Player it might look like this:

Streaming media being played by Windows Media Player.

Naturally your Raspberry Pi can also use its own web server. To play files on the same computer you should use the web address 127.0.0.1 (because it will be faster than trying to use the local computer via the network). Just run Midori and use the URL:

```
http://127.0.0.1:8080
```

MEAL PLANS
The Game of 'Snake'

Almost everyone has played a computer game, but have you ever thought how you might write one yourself? This section introduces the topic with a simple example: the game of 'Snake'. This was one of the simplest games on the BBC Micro, a computer which inspired Raspberry Pi. What game could be more appropriate for programming in Python?

Using Python for games
For many years it was believed that it was impractical to build high-performance games in an interpreted language like Python. Games usually have to work out the next thing to show before a computer display flickers on to its next frame. Since displays run at tens of frames per second a great deal of programming skill was required to do the work necessary before the deadline. Since then a number of things have improved the situation:

- Computers have increased in speed much faster than display frame rates.
- Graphical processing units, such as the one in Raspberry Pi, take away a lot of the work that a processor used to have to do.
- Interpreted languages have developed their own optimized libraries that perform much of the processor-intensive parts of a game in machine code.

Sadly the situation on Raspberry Pi doesn't benefit from all of these. It does have a fast processor and there are optimized games libraries in Python, but although, as we saw in the section about the processors in Raspberry Pi, there's a considerable quantity of GPU performance under the bonnet the GPU isn't easily accessible to Python (at least for now).

Nonetheless, quite elaborate games can be written on Raspberry Pi with quite a sophisticated look and feel.

Part of the reason this is possible is the provision of libraries, such as Pygame, where the logic of the game is written in Python but the handling of the display, sound, mouse and keyboard events etc are all implemented in the library.

In our example we'll illustrate the distinction between the game logic and the handling of the games platform by implementing 'Snake' once but then providing two 'back-ends': one that runs the game over a text console using a library called 'ncurses', and another that runs the game in a graphical window using 'pygame'.

Preparation
The previous sections about Python and programming languages in general might be of advantage to you if you haven't come across Python before.

The Python section described how Python programs can be developed using the 'Geany' integrated development environment and showed how it can be installed and used.

As indicated above, the programs to be written need

the 'ncurses' and 'pygame' libraries. The former is part of Python's standard library of modules and this command will install Pygame, if it isn't already:

```
sudo apt-get install python-pygame
```

Game description
This version of 'Snake' is played in a rectangular 'garden' that the eponymous snake can travel around at a certain speed in one of the four directions indicated by the arrow keys on a keyboard. At the edge of the garden there are barriers that will kill the snake if it hits them.

The other thing that the snake must not hit is itself. This is easy initially because the snake isn't very long, but it gets longer and longer as the game goes on, and faster and faster, making its own tail more difficult to avoid.

Somewhere in the garden there's a single target (probably a dead mouse). When the snake eats it (by hitting it) the snake grows in length by one unit and another dead mouse is thrown into the garden. The snake increases in speed and accumulates a score according to its length.

Game design
The graphics is greatly simplified by segmenting the snake into a series of identical body parts each of which occupies a single location on a grid that covers the garden. This means that the illusion of moving forward can be created by adding a body segment to the head of the snake at the same time as removing one from its tail. No movement or change in the intermediate body segments is necessary. Snake growth is simply managed by adding a body segment to its head but not removing one from its tail.

The program is to be split into two parts, each in separate Python source files. One part will deal with the game logic and the other will display the garden, the score and provide the events (keypresses) that drive the game.

Game logic
The game logic part ('snake.py') remembers the position of the snake in terms of coordinates on the grid and loops continually requesting an event from the game platform part ('board.py'). When it receives an event it'll be either a new direction for the snake to go in or a request to end the game; if no event occurs the previous event is reused (so that the snake continues in the same direction).

Once snake.py knows which direction to go in it asks board.py to sense what's in the next grid position in that direction. If it's free it asks the board to remove the snake's tail and adds its head. If it contains a target snake.py grows the snake by drawing its new head but not removing its tail, and it finds a location for a new target and asks board.py to display it. On the other hand if it's occupied by the barrier around the edge or by the snake itself the game is over.

When the length of the snake changes the score also changes and board.py is asked to display the new score.

Game platform

The description above leaves a number of jobs for the game platform to do: it must

- Create a garden and manage the placement of borders, snake segments and targets in terms of grid coordinates.
- Remove the tail segment of a snake.
- Add a head segment to a snake.
- Drop a target in the garden.
- indicate the next action provided by the user.
- Sense what's at a particular grid location.
- Display the score.
- Eventually delete the garden.

Each of these tasks will be accomplished by a different method (function) of a Python class called 'Board', which means that the board.py file will be expected to provide 'Board' and some names that indicate the possible actions and the possible things to sense.

In fact, we'll actually look at two different implementations: 'board_curses.py', which provides those items using the 'ncurses' library; and 'board_pygame.py', which provides the same things using the Pygame library.

Snake.py

This contains the main game logic. The following can be typed into Geany as it's described:

```
#!/usr/bin/env python
# -*- coding: utf-8 -*-
```

This program was created in Geany, which added these lines. The first line means that the resulting file can be made executable so as to appear like a command.

```
from random import randrange
```

This just obtains the 'randrange' function, which we'll use later, from the 'random' module.

```
# from board_pygame import *
from board_curses import *
```

The first line selects the 'pygame' version of the board and the second selects the 'ncurses' version. One of them should be commented out. The 'from' command will read the names defined in a Python file called 'board_curses.py' (or the compiled version, 'board_curses.pyc') and make them available to this program. We're expecting a class called 'Board' to be created and the constant variables starting 'ACTION_' and 'SENSE_' in the following code:

```
def snake_init(point, length):
    "Create a new snake with a tail at the
        given point"
    [x,y] = point
    return [[x+n, y] for n in range(length-1,-1,-1)]
```

We must remember the grid coordinates of each of the segments of a snake's body so that we can remember where they were when it comes time to delete them. To do this we use an array of points, where a point is a list of two items: its column and row. This function provides an initial value for a snake which has the given number of segments ('length') and whose tail is at 'point'.

The points in the list it produces are all in a line in front of the tail. A Python 'list comprehension' is used to create the list. It begins with a prototype list entry '[x+n, y]' (a point with the same row as 'point' and a column 'n' greater than 'point') and is followed by something that provides a number of instances of 'n' using the same syntax that's used in a 'for' loop. In this case n will take on the values starting at 'length-1' and ending at '0'. If 'point' was [10, 20] and 'length' was 3, for example, this would produce the list [[10+2, 20], [10+1, 20], [10+0, 20]].

The snake's head is the first item in the list and its tail is the last, so the snake's segments are in reverse order to the one in which they're read above.

The line of text on the second line of this function is optional in Python. If you do provide something here it'll be used when 'help' is used to enquire about the purpose of the function.

```
def find_target_pos(snake, width, height):
    "Find a spare board location for a random
        target"
    c = [place for tries in range(len(snake))
            for place in [[randrange(1,width-2,1),
                           randrange(1,height-2,1)]]
            if place not in snake]
    if c == []: # every try failed!
        target_pos = start_target
    else:
        target_pos = c[0]
    return target_pos
```

This function is a slightly odd way to go about what it does but it illustrates some other features of list comprehensions in Python. The one here is more complex and aims to try to find a random point on the board (but not on the edge where the barriers are) that isn't on the snake itself.

This time there are two variables that have many instances provided by the 'for' parts of the comprehension. The first 'tries' is simply a number from zero to the length of the snake less one. The second 'place' is generated for each instance of 'tries' and is taken from a list of one (and so is only ever given that one value). In this case the item in the list is point, whose column is a random number between 1 and width-2 and whose row is another random number between 1 and height-2. That is, the point is a random point that isn't on the outside of the grid.

The set comprehension doesn't include all of these values because it ends with an 'if' qualification. This means that only random points that don't occur in the snake will appear in the list.

The result will be, we hope, a number of candidate points to place the target. If the list of candidates is empty it'll mean that all of the points that were chosen at random were in the snake. In this case we're very unlucky and we arbitrarily choose the snake's original starting point. Otherwise the first candidate in the list is taken and returned.

```
def main():
    "Play the 'snake' game"
```

This defines the function that'll be called to play snake:

```
board = Board()
(board_width, board_height) =
    board.size()
```

First we obtain an instance of each of the methods and attributes in the Board class defined in whichever board implementation we chose. We use the 'size' method to discover how wide and how high the garden grid is that we've been provided with.

This uses the feature of Python that allows a number of variables to be set at once from a list or a tuple. Each of the values from the list on the right-hand side of the '=' are assigned in turn to each of the variables in the tuple of variables on the left-hand side.

```
start_len = 6
start_pos = [(board_width-start_len)//2,
    board_height//2]
snake = snake_init(start_pos, start_len)
```

We choose a snake length of six, a starting position in the middle of the board and then use the 'snake_init' function discussed above to create the initial representation of the snake's segments.

The '//' operator above is similar to the normal division operator '/'. For example, both 8/2 and 8//2 will give 4. The difference is that the result is always a whole number with '//' but may have a fractional part with '/'. So 9/2 is 4.5 but 9//2 is 4 (the fractional part is simply discarded). It's used here because it doesn't really make sense to have fractional grid positions.

```
start_target = [board_width//4*3,
    board_height//4]
board.place_target(start_target)
```

This, fairly arbitrarily, selects a place for the first target and places it on the board using its 'place_target' method.

```
action = ACTION_RIGHT
```

This action is the one that'll be used if the user doesn't provide anything to do.

```
score = -1 # impossible score
```

Setting the score to an impossible value means that it'll be updated immediately once it's checked.

```
while action != ACTION_STOP:
```

The program goes around this loop once every time interval and usually moves the snake by one position. It'll continue to do this until an action to stop the program has been set.

```
if score != len(snake)-start_len:
    score = len(snake)-start_len
    board.show_score(score)
```

We're using the number of segments that have been added to the snake so far in the game as the score. If the number of segments has changed we update the score and tell the board to re-display the new score.

```
next_action = board.get_action(180 -
    score*3)
if next_action != ACTION_NONE:
    action = next_action
```

Next the user is given a short while to cause an action to be reported. The 'get_action' method is given the exact amount of time that we'll wait before going on. When we do go on it'll be to move the snake forward, so the amount of time we wait here effectively determines the speed of the snake. The strange formula for the amount of time starts by waiting 180 thousandths of a second but varies as the score increases, reducing the wait by three for every extra score. This means that there'll be no wait at all by the time the snake is 60 segments long.

If the user hasn't produced an event (pressed a key) the previous action is repeated. On the other hand if the action was 'ACTION_STOP' this will be seen when we start the loop again and the program will then come to a halt.

```
[x,y] = snake[0]
if action == ACTION_RIGHT: x = x+1
elif action == ACTION_LEFT: x = x-1
elif action == ACTION_UP: y = y-1
elif action == ACTION_DOWN: y = y+1
board.snake_unplot_tail(snake)
snake.insert(0, [x, y]) # insert point
    at snake[0]
```

Here the coordinates of the head of the snake are taken and used to work out the new coordinates of the head for the moved snake. The new head location is placed at the start of the 'snake' list of grid coordinates but the display isn't updated yet.

```
sense = board.sense(snake[0])
if sense == SENSE_NONE:
    snake.pop() # forget the tail
    board.snake_plot_head(snake)
elif sense == SENSE_TARGET:
    target = find_target_pos(snake,
        board_width, board_height)
    board.place_target(target)
    # don't forget the old tail
        location, get longer instead
    board.snake_plot_head(snake)
else:
    action = ACTION_STOP
```

Depending on what the board says is found at the new location for the snake's head (snake[0]) the code will do one of three things. If there's nothing there it forgets the location of the snake's tail and draws the new head on the board. This will provide the illusion of the snake moving on one location. If a target (dead mouse) was there the tail isn't forgotten, nor is it removed from the board, which will make the snake appear to move on by one location while also becoming longer. In due course the new length will be reflected in a higher score and result in higher speed too. Because we'll have written over the target the code doesn't bother erasing it specifically but it does use the 'find_target_pos' function we discussed above, to retrieve a new location for it and plot it in its new position.

If something's been sensed that wasn't the target the code assumes that it was a barrier or the snake itself. In this case the code pretends that a stop action was requested.

```
board.end()
```

Once the loop has finished, the implementation of the board is told that its job is done. We print the final score to the screen (not the board) and exit with a return code of zero (which indicates that everything has been successful).

```
print "Final score: "+str(len(snake)-
    start_len-1)+'\n'

    return 0
```

```
if __name__ == '__main__':
    main()
```

If this file was called as a module 'main' won't be called but the functions defined in 'snake.py' will be available to the program that included it.

Board_curses.py

The task of this code is to create a version of Board which creates the snake's garden with the help of the 'curses' library. This library allows a section of a terminal screen (which it calls a 'windows') to be reserved for reading and writing individual characters to. The library doesn't allow arbitrary shapes to be written to the windows: terminal screens can only display terminal screen characters, one in each character position; but they do allow characters to be written to any position in the window.

The program can live within these restrictions by representing the segments of a snake's body as 'X' characters and the target as a '0' character, with various other characters being used for the barrier around the edge of the garden.

Line-by-line

```
#!/usr/bin/env python
# -*- coding: utf-8 -*-
#

width = 2+78
height = 2+22
```

Rather than using the numbers 80 and 24 in the code (because we're going to use windows that are 80 characters wide by 24 tall) we use these two variables to hold these values. Their names will make it more obvious what we mean.

```
from curses import initscr, curs_set, newwin,
  endwin, KEY_RIGHT, \
              KEY_LEFT, KEY_DOWN, KEY_UP
```

Instead of importing every name provided by the 'curses' library this code lists each one explicitly.

Next we provide convenient values for the symbols that Snake uses for actions and sensed objects.

```
ACTION_NONE = -1
ACTION_LEFT = KEY_LEFT
ACTION_RIGHT = KEY_RIGHT
ACTION_DOWN = KEY_DOWN
ACTION_UP = KEY_UP
ACTION_STOP = 27

SENSE_NONE = 32
SENSE_TARGET = ord('0')
SENSE_SNAKE = ord('X')
```

A Python class is like a special dictionary of values. As well as normal Python values there are special function values, called 'methods', that can be named. When an instance of a class is created, for example like this,

```
board = Board()
```

a new copy of the dictionary in Board is assigned to the class instance ('board'). Values in this dictionary are accessed by name using a full stop after the name of the instance. For example, if the class 'Board' included a method 'size', say, it could be invoked using

```
board.size()
```

Although not specified explicitly in this notation the first argument to every method will be the instance's dictionary. Typically method definitions call this formal parameter 'self'.

There's much else to be said about classes in Python and in other languages that we don't address here.

The Board class is one of the things that snake.py expects this file to define.

```
class Board:

    def __init__(self):
```

When an instance is taken of class arguments the '__init__' method is implicitly called. That is

```
board = Board()
```

is similar to

```
board = Board.__init__()
```

Typically a classes '__init__' method will provide initial values for names in the instance's dictionary ('self') and that's what's done here with 'self.xlen', 'self.ylen' and 'self.win'.

```
initscr()
```

clears the screen, and puts the terminal into a special state and readies the library for use.

```
curs_set(0)
wx = width
wy = height
self.xlen = wx
self.ylen = wy
```

makes the cursor invisible and remembers the chosen size of the screen, which size is then passed on to the curses library to create a new window class instance:

```
win = newwin(wy,wx,0,0)
```

As an instance of a class, 'win' will have a number of methods defined on it. For example, 'win.border()' and 'win.nodelay()', which are used below.

```
win.keypad(1)
```

allows the arrow keys to be interpreted as single characters (although, in reality, those keys often generate a short sequence of characters beginning with an 'escape' character).

```
win.nodelay(1)
```

allows requests for a character typed into the window to be answered by 'nothing was typed' instead of waiting until a key is pressed.

```
win.border('|','|','-','-','+','+','+','+')
self.win = win
```

draws a boundary around the edge of the window made up from the characters supplied for its different sections – all of which will have prepared the initial state of the snake's 'garden'.

```
def size(self):
    return self.xlen,self.ylen
```

This method simply returns the size of the window area we've prepared to the caller.

```
def end(self):
    endwin()
```

When the game has finished this method calls 'endwin' in the curses library, which returns the terminal to its normal state.

```
def get_action(self, delay):
    self.win.timeout(delay)
    return self.win.getch()
```

Using two window methods this function will wait for the required delay and then read any character that might have been typed during that time. Because we've used 'win.nodelay(1)' this will return -1 if no character was typed. The values for the variables starting 'ACTION_' above have been chosen carefully to match the values that'll be returned if the arrow keys are used.

```
def place_target(self, point):
    self.win.addch(point[1], point[0], '0')
```

The window 'addch' method places a character at the given row and column of the grid of characters in the window. Placing the target is implemented by writing a '0' at the given coordinates.

```
def sense(self, point):
    return self.win.inch(point[1],
    point[0]) & 255
```

The windows class also provides a method 'inch' which will read the character that currently occupies the specified grid coordinates. The values of the variables beginning 'SENSE_' above have been chosen to correspond to the characters that the program might find on the board.

```
def show_score(self, score):
    self.win.addstr(0,2,' Score:
    '+str(score)+' ')
```

Just as 'addch' writes a single character at the given window location, 'addstr' writes an entire string. We write the text showing what the score is at a specific point on the boundary so that it'll overwrite the last score that was written there.

```
def snake_unplot_tail(self, snake):
    self.win.addch(snake[-1][1],snake[-1]
    [0],' ')
```

The 'addch' method is used again but this time to over-write whatever character used to be at the given coordinates with a space character, which will make the previous character seem to disappear.

```
def snake_plot_head(self, snake):
    self.win.addch(snake[0][1], snake[0]
    [0], 'X')
```

Plotting the head is achieved by writing an 'X' to the relevant grid coordinates.

Running the program
Now that you have both the program logic in snake.py and a board implementation in board_curses.py you can try running it. Make sure that both files have been saved.

If you're using Geany select the file snake.py and make sure the first few lines still contain

```
# from board_pygame import *
from board_curses import *
```

Then click on the 'execute' button to start the game.

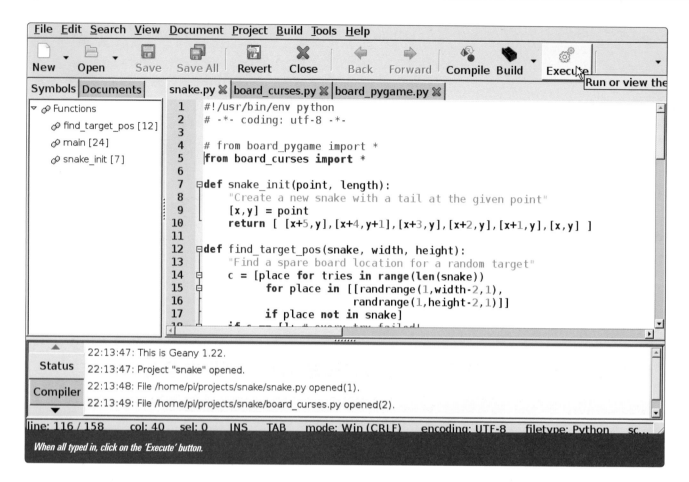

File Edit Search View Document Project Build Tools Help

New | Open | Save | Save All | Revert | Close | Back | Forward | Compile | Build | Execute

Run or view the

Symbols | Documents

snake.py | board_curses.py | board_pygame.py

Functions
- find_target_pos [12]
- main [24]
- snake_init [7]

```python
1   #!/usr/bin/env python
2   # -*- coding: utf-8 -*-
3
4   # from board_pygame import *
5   from board_curses import *
6
7   def snake_init(point, length):
8       "Create a new snake with a tail at the given point"
9       [x,y] = point
10      return [ [x+5,y],[x+4,y+1],[x+3,y],[x+2,y],[x+1,y],[x,y] ]
11
12  def find_target_pos(snake, width, height):
13      "Find a spare board location for a random target"
14      c = [place for tries in range(len(snake))
15          for place in [[randrange(1,width-2,1),
16                         randrange(1,height-2,1)]]
17          if place not in snake]
```

Status

22:13:47: This is Geany 1.22.
22:13:47: Project "snake" opened.

Compiler

22:13:48: File /home/pi/projects/snake/snake.py opened(1).
22:13:49: File /home/pi/projects/snake/board_curses.py opened(2).

line: 116 / 158   col: 40   sel: 0   INS   TAB   mode: Win (CRLF)   encoding: UTF-8   filetype: Python   SC...

When all typed in, click on the 'Execute' button.

You should see a window like this appear with a snake moving as you direct it with the arrow keys on your keyboard.

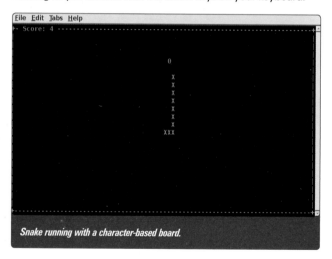

Snake running with a character-based board.

Board_Pygame.py

Although the game implemented above works it looks rather drab. The next board, which uses Python's Pygame library, will make it more colourful. It won't really illustrate the full potential of this library, but we'll go on to discuss the possibilities once we illustrate the simple case.

To Pygame the visuals in a game consist of a series of new pictures, or 'frames', displayed one after another on a display. A typical game in Pygame will have a target rate at which new

frames are produced. It takes a little time to update the display to show a new frame and if the frame being displayed changes during that time the display will look as if it has one or more 'tears' in it, so a typical game prepares a picture to be displayed and then hands it on to a separate part of the system that will schedule it for display without further modification.

Displays can show only a certain number of frames in a second (perhaps 25 or 30) and if new frames are produced more often some will have to be dropped. Consequently games often have a notion of a rate at which they aim to produce new frames. Pygame provides a special 'clock' that can be used to manage the display of new frames.

The way this board works is to maintain a pixel-by-pixel copy of the board we want to display in a Pygame 'surface' (which is provided for just this use) and to accumulate a list of areas of the surface that'll need updating to change the current frame into the next, so that the basic structure of the game is expected to be something like

```python
display_frame()
while not_finished():
    changes = update_frame()
    wait_for_next_frame_clock()
    display_frame_parts(changes)
```

(This isn't the code we actually use, just an illustration of how it'll work.)

The changes are made on the game surface, but those new pixel patterns aren't displayed until the display is ready for the next frame. An individual change is just a rectangular area of the surface that needs redrawing. Pygame provides a special object called a 'Rect' to represent such rectangles.

Although many games have frame rates of many times a second our aim in snake.py is only to update the screen every time we need to move the snake. Updating tens of times per second would make the game unplayable because the snake would move too quickly.

Line-by-line

Like the curses version of the board, we start by calling in some libraries, defining some variables that'll be important in configuring the code, and providing values for the 'ACTION_' and 'SENSE_' values that snake.py uses.

In this case the most important library is 'pygame' which is actually a collection of component libraries, some dealing with different peripherals such as mice, cd-roms, joysticks and MIDI music synthesisers; some dealing with the surfaces to draw into and the fonts, sprites and shapes that you might like to place on them; some dealing with music and video playback; and others looking after the handling of events that occur as the game unfolds.

```
#!/usr/bin/env python
# -*- coding: utf-8 -*-
#

import pygame
from pygame.locals import *
from random import randrange

SCREEN_PX = (800, 640)
SCALE = 20  # pixels per square

ACTION_NONE = -1
ACTION_LEFT = K_LEFT
ACTION_RIGHT = K_RIGHT
ACTION_DOWN = K_DOWN
ACTION_UP = K_UP
ACTION_STOP = K_ESCAPE

SENSE_NONE = 0
SENSE_TARGET = 1
SENSE_SNAKE = 2
SENSE_BARRIER = 3
```

As in the curses example, the first thing we define is the 'Board' class that snake.py expects, and the '__init__' function that initializes it.

```
class Board:

    def __init__(self):
        pygame.init()
```

This call is necessary before any other calls can be made to the Pygame library.

```
        width, depth = SCREEN_PX
        status_depth = 2*SCALE
        self.clock = pygame.time.Clock()
```

This is the clock that'll be used to wait for the next frame time.

```
        self.screen = pygame.display.set_
            mode(SCREEN_PX)
        pygame.display.set_caption('Snake')
```

This creates a new surface with the given size and gives it a title 'Snake'. It will appear in a window of its own.

```
        self.status_depth = status_depth
        self.garden_rect = Rect((0,0), (width,
            depth-status_depth))
        self.status_rect = Rect((0, depth-
            status_depth),
                    (width, status_depth))
        self.gardensize = (width//SCALE,
            (depth-status_depth)//SCALE)
```

Here we're recording the size and shape of the display defining two rectangles that completely fill the surface, one taking the majority of the space for the garden (garden_rect) and another at the foot of the surface where the score will be displayed (status_rect).

```
        self.back_col = (220, 10, 10) #
            background colour
        self.statusback_col = (20, 20, 180) #
            status background colour
        self.snake_col = (0,240,90) # snake
            colour
        self.barrier_col = (0,0,0) # barrier
            colour
        self.target_col = (10,10,200) # target
            colour
        self.score_col = (250,210,0) # score
            font colour
        self.score_font = pygame.font.
            Font(None, 36)
        if self.score_font == None: print
            ("No font found for the score")
```

The above sets some variables that'll determine the look of the board, with colours for various items and the font to use when displaying the score. Colours in Pygame consist of a tuple of three numbers between 0 and 255: the first represents the quantity of red in the colour, the next the quantity of green and the last the quantity of blue. The colour (0,0,0) is black and the colour (255,255,255) is white.

```
        self.barriers = []  # list of barrier
                            rectangles
        self.targets = []   # list of target
                            rectangles
        self.snake = []     # last seen snake
                            (list of board coords)
        self.updates = []   # screen areas we
                            haven't updated yet
```

These are the main variables that we'll use to keep track of objects on the surface.

```
    self.reset_screen()
```

Clear the screen to finish our initialization (this function is one that we define a little further on).

```
def draw_barriers(self, width):
    topleft = self.garden_rect.topleft
    gwidth, gdepth = self.garden_rect.size
    rects = [pygame.draw.rect(self.screen,
        self.barrier_col, Rect(topleft,
        (width, gdepth))),
            pygame.draw.rect(self.screen,
            self.barrier_col,
            Rect((gwidth-width,0),
            (width, gdepth))),
            pygame.draw.rect(self.screen,
            self.barrier_col,
            Rect((width,0),
            (gwidth-2*width, width))),
            pygame.draw.rect(self.screen,
            self.barrier_col,
            Rect((width,gdepth-width),
            (gwidth-2*width, width)))
    ]
    return rects
```

This function uses the 'rect' function in pygame.draw to create rectangles on the surface (self.screen) that represent the barriers. Each call to this function returns a Rect object and the list of these rectangles is returned to the caller. In principle, if we wanted to modify the game to produce a more elaborate set of rectangles (perhaps in successive levels of the game) this is the function that should be updated. At the moment four rectangles are used to create a barrier around the outside of the garden.

```
def reset_screen(self):
    "Wipe the screen"
    self.screen.fill(self.back_col)
    self.barriers = self.draw_barriers(SCALE)
    pygame.draw.rect(self.screen, self.
        statusback_col, self.status_rect);
    pygame.display.update(self.garden_rect)
```

To reset the screen we'll fill it with pixels of the background colour, draw the barriers around the garden, fill the status area with its background colour and then pass the surface we've written to the display to be shown.

```
def pix_pos(self, point):
    "Return the pixel position of board
        coordinates"
    return (point[0]*SCALE, point[1]*SCALE)
```

The garden is imagined to be a grid by snake.py, but each of the squares in a grid will use a number of pixels. This function converts the grid coordinate to the relevant square's pixel coordinates. A point is represented as a tuple of two numbers representing a column and a row, so 'point[0]' will be the column and 'point[1]' will be the row.

```
def plot_blob(self, point, colour):
    "Plot a coloured blob as part of a
        snake or as a target etc"
    pix_x, pix_y = self.pix_pos(point)
    rect = pygame.draw.circle(self.
        screen, colour, (pix_x+SCALE//2,
        pix_y+SCALE//2), SCALE//2)
    self.updates.append(rect)
    return rect
```

This version of the board makes a simplifying assumption that parts of the snake's body and the target (food) can be represented by a small circle (blob) of different colours. This routine draws one on the screen's surface and returns the rectangle that bounds it.

Most of the drawing functions in Pygame work like this. They make a change to the pixels on the surface and then return a rectangle object for the smallest rectangle that can be drawn around the changes (a bounding rectangle).

```
def size(self):
    "Return the size of the screen in
        board coordinates"
    return self.gardensize
```

Notice that this returns the size of the garden section of the board, not the whole screen (it doesn't include the status area at the bottom).

```
def end(self):
    "Discard the display at the end of
        the game"
    pygame.display.quit()
```

When the game is finished this Pygame function is called to remove the window that's been being used for the display.

```
def get_action(self, delay_ms):
```

This function is supposed to wait for something to happen and then tell snake.py what it was. We're going to treat this delay as the wait for the next game frame and an opportunity to display the one that we've just worked out.

```
if len(self.updates) > 0:
    # update the indicated areas of
        the screen
    pygame.display.update(self.updates)
    self.updates = []
```

Updates all the parts of the display whose bounding rectangles are in the self.updates list and then empties the list ready for the next frame.

```
    # wait for the next frame
    self.clock.tick(1000/delay_ms)
```

Wait a duration that will bring us to the next frame time. During this time Pygame will accumulate a list of events for each of the things that happens.

```
        action = ACTION_NONE
    for event in pygame.event.get():
        newaction = action
        if event.type == pygame.QUIT:
            newaction = ACTION_STOP
        elif event.type == KEYDOWN:
            if event.key in (K_ESCAPE,
                            K_LEFT, K_RIGHT,
                            K_UP, K_DOWN):
                newaction = event.key
        if action != ACTION_STOP and action
                    != K_ESCAPE:
            action = newaction
    return action
```

Go through the list of events (if there are any) to work out which event to tell snake.py about. We're interested only in two types of events: the STOP event (caused if the user deletes the game window, for example); and a key press event. Pygame can supply a lot of other events too (such as those caused by the mouse, for example) but we ignore these. If the game was to be updated to use the mouse or a joystick to steer the snake this is the function we'd change.

```
def place_target(self, point):
    "Add a new target for the snake to
    find"
    self.targets = [self.plot_blob(point,
    self.target_col)]
    self.updates = self.updates + self.
    targets
```

Here we use our routine to draw a blob to update the surface with the picture of a target (snake food). We remember a list of target places and add these on to the list of locations we need to update when we draw the next frame.

We don't really need to remember a list of targets because we only ever plot one. However, if we updated snake.py to create several targets instead of just one this representation would be quite useful.

Although we just plot a blob there's no reason why this routine couldn't draw a tiny picture of the snake food in question. It wouldn't be difficult to have several and choose one at random (frog, baby bird, dead mouse etc).

```
def sense(self, point):
    "Return an indication of what is found
    at board coordinates"
    if point in self.snake[2:]: # we've
    hit ourself!
        return SENSE_SNAKE
    else:
        point_rect = Rect(self.pix_
        pos(point), (SCALE, SCALE))
        if point_rect.collidelist(self.
        targets) >= 0:
            return SENSE_TARGET
        elif point_rect.collidelist(self.
        barriers) >= 0:
```

```
            return SENSE_BARRIER
        else:
            return SENSE_NONE
```

This uses the Pygame Rect member function 'collidelist'. It takes a list of other rectangles and indicates which one of them intersects ours. If none intersect ('collide' with) ours the function returns -1.

After checking that the point being sensed (which is the snake's head) isn't one of the coordinates in the rest of the snake this function checks that it doesn't collide with either one of the targets or one of the barrier rectangles. If it does collide with one of these the function returns the appropriate 'SENSE_' value.

```
def show_score(self, score):
    upd = pygame.draw.rect(self.screen,
    self.statusback_col,self.status_rect);
    # Make a mini-surface containing the
    anti-aliased score text
    score_text = self.score_font.render
    (' Score: '+str(score),True, self.
    score_col)
    textpos = score_text.get_rect()
    # put the centre of the text in the
    centre of the status area
    textpos.centerx = self.status_rect.
    centerx
    textpos.centery = self.status_rect.
    centery
    self.screen.blit(score_text, textpos)
    pygame.display.update(upd)
```

To show the score we first create a new small surface containing only the letters of what we want to show as text (using the render() member function of the font object we're using). We then do something a little clever: we retrieve the Rect object that represents the small surface's bounding rectangle and update its details so that it describes the same rectangle but centred in the centre of the score area. We then use that updated position to determine where on the screen surface to write the surface containing the text.

The function we use to write the text surface (which is composed of pixels) on to the main surface (also pixels) is called 'blit' and can be used to draw any surface on to any other.

```
def snake_unplot_tail(self, snake):
    self.plot_blob(snake[-1],
    self.back_col)
```

To unplot the tail we simply draw a blob the colour of the background in the last snake coordinate (snake[-1] will be the grid coordinates of the last segment of the snake).

Naturally this routine would have to be more complex if the colour of the background wasn't the same everywhere.

If we'd used a different shape for the snake segments, or for the tail segment in particular, this is the routine we'd have to follow to update this function.

```
def snake_plot_head(self, snake):
    self.plot_blob(snake[0],
      self.snake_col)
    self.snake = snake # remember the
      last snake
```

Very much like the function that unplots the tail, this function simply plots a blob where the new head of the snake is supposed to be, but in the colour used for the snake's body. In addition it remembers the whole of the snake (so we can use it in the 'sense' function to work out whether we've hit ourself).

Because we have the coordinates for the whole snake we could update any or all of the segments in this routine. Updating the head is the least work it could do.

We could make the snake's tail or head a different shape to the other segments by updating the location of the old head and new tail as well as setting the new one.

To make the snake into a centipede we might draw sprites for each body segment with legs. If we alternated the leg sprite we used we could make the centipede appear to walk. With a little more work we could make it appear to wiggle from side to side as it progressed.

Running the program

Now that you have both the program logic in snake.py and a board implementation in board_curses.py you can try running it. Make sure that both files have been saved.

If you're using Geany select the file 'snake.py' and make sure the first few lines still contain

```
# from board_pygame import *
from board_curses import *
```

then click on the 'execute' button to start the game.

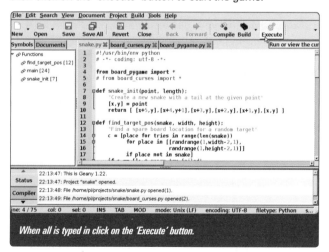

When all is typed in click on the 'Execute' button.

This time you should see a window like this appear:

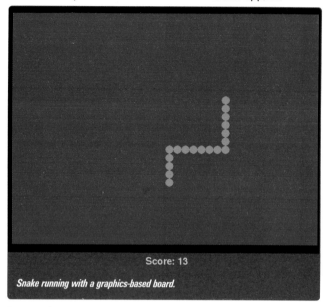

Score: 13

Snake running with a graphics-based board.

Interfaces and abstraction

The way this game was designed separated two parts: the part that dealt with the logic of the game and the part that dealt with its representation on a display, of which we built two versions.

This was possible because the features that the first part needed of the second were the same. In this case the details that needed to be established were the methods of a Python class that needed to be provided and the names of some constant values to represent directions and actions. These details make up the 'interface' between the two parts of the program.

Designing complicated tasks by dividing them up into parts with specific interfaces between them is a basic feature of good program design. Once the interfaces have been specified properly, each part of the program as a whole could, in principle, be built by different people. For example, once the snake-board interface has been established one person could have written the game logic, another could have written the ncurses board and yet another could have written the Pygame board.

Specifying an interface can be a tricky task. If some details that'll be needed by the game aren't included the board designers would build something that won't quite be able to run the game. If too many details about the design are included (for example, specifying an image to be used for the graphical elements of the board) the range of boards will be cut down (the ncurses board couldn't be built). Just 'enough' details must be included and no more. The interface shouldn't include all the details of the implementation, some should be taken away (abstracted).

The amount of detail in the interface is sometimes referred to as its 'level of abstraction', which must be neither too little nor too great; and the process of deriving an interface from an envisaged implementation is called 'abstraction'.

Other examples

There are several other games all programmed in Python using Pygame in the standard distribution in the directory

```
/home/pi/python_games
```

MEAL PLANS
Twitter Alert

A twitter alert is a device that requires both Internet access and some simple hardware. The idea is to activate an electronic toy when a chosen phrase is spotted in the twitter-sphere.

Yoda Twitter.

Hardware

To build the alert we'll use a Pi-Face interface board (to both protect the Raspberry Pi against electrical miswiring and to provide a convenient relay switch) and a battery-operated toy. Different battery-operated toys will, of course, provide different aesthetics. Some toys will move when a button of some kind is depressed, others make a noise, and others simply provide a light display. The one in this example is a talking Yoda toy which will repeat one of three or four phrases when a button is pressed, including 'Do. Or do not. … There is no "try".'

Pi-Face

The Pi-Face interface board can be obtained from the School of Computer Sciences at Manchester University at

```
http://shop.openlx.org.uk/
```

Although it provides eight GPIO inputs and eight GPIO outputs these aren't the GPIO lines provided natively in Raspberry Pi. Instead they're driven via a separate chip (ULN2803A) which is controlled using one of Raspberry Pi's Serial Peripheral Interfaces (which are described in the section about SPI).

We discussed the fragility of the GPIO lines connected directly to Raspberry Pi in the section about GPIO lines. The inputs on Pi-Face can easily deal with 5V signals and the board producers allow an output current of up to 200mA (both of these ratings are comfortably within very wide tolerances).

The outputs have accompanying LEDs (eight of them in the current version of the board and four in an older one), and four of the inputs have associated push-button switches. Of most interest to this project, however, is the attachment of two of the outputs to very capable relays which, in principle, can switch large voltages at up to 10A.

Hardware preparation

Yoda

Like many toys, the electronics in Yoda aren't intended to be accessible. In order to find the switch (which is also the main printed circuit board for the electronics) it's necessary to undertake an abdominal procedure using a sharp pair of scissors. Luckily the scar won't show in this case because Yoda is wearing a tunic that can be dropped back down over the incision later. The first step is to expose Yoda's belly button, trying to keep to the moderate practices of keyhole surgery. The button is a conductive pad that's pushed over a sinuous area of the circuit board that looks like the interlocking fingers of two hands. Two wires should be connected to this area, with one wire connected to a finger from one of the 'hands' and another connected to the other. The easiest way to do this is to use a soldering iron.

Pi-Face Raspberry Pi interface board.

The first incision.

Removing the tummy button.

Wires for an external button.

After the tummy tuck.

Connecting Yoda to Pi-Face.

You should be able to test how successful you've been in avoiding a short circuit by touching together the other end of the two wires to make sure the toy still behaves as it should when the button is pressed.

Pi-Face

Pi-Face is designed to fit directly on top of a Raspberry Pi board using the expansion header as both an electrical and a mechanical connector.

To attach Pi-Face simply position the expansion connector pins under the corresponding socket on Pi-Face and press the two together.

Pi-Face connected to Raspberry Pi.

To connect Yoda take the two wires from the toy and, using a screwdriver, attach them to the left-hand pair of holes in the grey connection block in the centre of the rear of Pi-Face.

Software preparation

SPI device

To use Pi-Face you'll need to set up SPI on Raspberry Pi as described in the section about SPI. In particular you need to

- Ensure that the spi_bcm2708 module loads.
- Make sure 'udev' gives the SPI devices usable access controls.

Pi-Face Python library

To use Raspberry Pi we'll make use of the Python library that the School of Computer Sciences at Manchester University has already provided. This is contained in a software release that's provided using the 'git' source control program (which we discussed briefly in the section about recompiling the Linux kernel).

To ensure that you have this and other useful programs use

```
sudo apt-get install git make gcc python-dev
 python-gtk2-dev
```

You'll need somewhere to place this software. One suggestion is to keep it alongside any other source trees you acquire in a subdirectory of your home directory called 'tree' (but you can place it where you like).

```
mkdir -p ~/tree/piface
cd ~/tree/piface
git clone https://github.com/thomasmacpherson/
 piface.git
```

This creates a place to put the files, selects it as your current working directory and then uses the 'git' command (installed in the previous step) to make a copy of the programs. (Because this code is still being developed you may find that you'll get access to improvements in the code at a later date by executing the command

```
git pull
```

in this directory.)

You should find that a directory called 'piface' has been created. To install the Python software you'll need to use

```
cd piface/python/
sudo python setup.py install
```

This is all we need from this software at the moment, but do have a look at the 'README.md' file which details the other features that it provides, including a C library and an 'emulator' which will represent the effects of any program that you execute graphically.

Twitter Python library
The Python library that supports access to your Twitter account is provided on the Web at

```
http://code.google.com/p/python-twitter/
```

It requires one or two Python libraries before it'll work properly, which you can install using

```
sudo apt-get install python-simplejson
 python-httplib2  python-oauth2
```

The library itself is obtained from this website:

```
http://code.google.com/p/python-twitter/
 downloads/list
```

It's updated with different versions from time to time. If the most up-to-date version is 0.8.2 you could use 'wget' to obtain a copy:

```
mkdir -p ~/tree/twitter
cd ~/tree/twitter
wget http://python-twitter.googlecode.com/
 files/python-twitter-0.8.2.tar.gz
```

(The wget command should all be typed on the same line.) This will create a file 'python-twitter-0.8.2.tar.gz' which can then be unpacked using

```
tar -zxf python-twitter-0.8.2.tar.gz
```

which should result in the directory 'python-twitter-0.8.2' appearing. The Python library can now be installed using

```
cd python-twitter-0.8.2
python setup.py build
sudo python setup.py install
```

This will be sufficient for our needs. You may be interested, however, in the range of things that this Twitter software can do for you. They're described at

```
http://code.google.com/p/python-
 twitter/
```

which has examples of the way that you can get a list of your friends, find the most recently posted texts by those you're following and even make a new tweet.

The documentation for all the functions that Twitter make available are documented at

```
http://dev.twitter.com/doc
```

Python development environment
If you haven't done so already, you could install a program to use for developing Python programs. We describe how to use 'Geany' in the section about Python. It's installed using

```
sudo apt-get install geany
```

Test the toy
First we should write a quick Python program that'll demonstrate whether or not our hardware is working. You can do this by creating a new Geany project and typing this program into a new Python file 'yoda.py'.

You should see or hear your toy perform its action when the 'Execute' button is used. (In the example, the toy said 'A Jedi uses the force.')

Naturally this Python program can be used as it is for various purposes in other programs. For example, it could be incorporated into a Python or Bash program to indicate an error condition. If, like this toy, your toy makes a noise it could be made the subject of a cron entry to turn it into a daily alarm (see the section about running programs regularly).

Scan the world's tweets

Adding a few more lines will give the program the ability to scan tweets looking for an important word or phrase. In this case we use 'yoda' and the result is that the toy gives one of its pre-set exclamations every time anyone in the world tweets a message involving that word (which turns out to be one or two times per minute).

This is the example program:

```
# use the Pi-Face input and output library
import piface.pfio as pfio
# get the twitter library
import twitter
# use time to sleep
import time
class Toy(pfio.Relay):

    def __init__(self, relay_number):
        pfio.Relay.__init__(self, relay_number)

    def pulse(self, seconds):
        self.turn_on()
        time.sleep(seconds)
        self.turn_off()
```

This is the same class that we used in our test. The number of the relay to use on the Pi-Face board is given when an instance of it is created. It provides one method 'pulse' which turns the relay on, waits a number of seconds, and then turns it off again.

```
class Twit():

    def __init__(self, mentioning):
        self.me_on_twitter = twitter.Api()
        self.last_status = twitter.Status()
        self.mentioning = mentioning

    def new_mention(self):
        """
        Return True if a new Twitter status
         containing our mention
        has arrived
        """
        status = self.me_on_twitter.
         GetSearch(term=self.mentioning,
          per_page=1)[0]
        # print(status)
        if status.id != self.last_status.id:
            self.last_status = status
            print("%s says: %s"%(status.user.
             screen_name, status.text))
            return True
        else:
            return False
```

When a new instance of the Twit class is created it remembers a phrase, 'mention', to seek. It also provides only one method, new_mention, which returns 'True' and prints out the tweet whenever the latest search for the phrase we're looking for gives a new result.

The object returned by Twitter's GetSearch method is a dictionary object with

Field	Type	
created_at	String	The date the tweet was made
id	String	A unique 18-digit number
source	String	The name of the program producing the tweet
text	String	The text of the tweet
user	Dictionary	Containing the following:
profile_image_url	String	The URL of a picture of the tweeter
screen_name	String	The tweeter's ID

```
def main():
    pfio.init()
    yoda = Toy(0)
    tweets = Twit("yoda")
    poll_sec = 2 # number of seconds between
     twitter status requests
    while True:
        if tweets.new_mention():
            yoda.pulse(1)
            time.sleep(poll_sec-1)
        else:
            time.sleep(poll_sec)
    return 0

if __name__ == '__main__':
    main()
```

The main program will initialize the Pi-Face input/output library and create a Toy object, called Yoda, on relay number 0. It also creates a Twit object that searches for the phrase 'yoda' and then loops searching for a new mention of that phrase. When one is found the toy is given a one-second pulse and then waits the rest of the two seconds we're using as our poll interval. If none are found we simply wait two seconds (and then check again).

Do more

It wouldn't be difficult to change this program in any number of ways: adding more hardware to sense or to activate (there are already plenty of LEDs and switches unused on Pi-Face); adding more Twitter events to look for (such as tweets to yourself, or activity by any of your friends); or doing more with the tweet information (such as accumulating the relative number of tweets from Android devices, iPhones and PCs). You could involve additional software to make information available on your own Raspberry Pi website, send a tweet or send an email.

Raspberry Pi can use the 'omxplayer' command to play either audio or video files, and that can be called by Python. A short burst of a *Star Wars* preview could probably be played for relevant tweets, with a little ingenuity.

The Pi-Face documentation (which already has a very similar example, but involving chickens) has lots of other ideas for projects and shows how it's possible to speak the contents of tweets quite simply from Python.

MEAL PLANS
A Media Centre

G aming on a Microsoft Xbox has always been fun, but what do you do when your Xbox, or Xbox 360, is replaced by something even more exciting? The people who designed 'XBMC' had the answer to this: reuse it as an XBox media centre. The program is open source (as discussed in the introduction) and liberally licensed under the GNU General Public License. The resulting program has been so successful that it's no longer tied to Microsoft operating systems and now runs on many Unix-based operating systems including Mac OSX, Apple iOS, Android and Linux.

As a media centre, XBMC can organize and play music, video and still photos. But, partly because it supports a wide range of extensions, or 'plugins', it's also capable of much else, including various widgets, web interfaces, weather forecasting, themes and 'web scrapers' (which extract information from web pages). One of the good things about XBMC for Raspberry Pi users is that these plugins are written in Python, and new user-supplied plugins are always welcome.

As we discussed earlier, Raspberry Pi can be connected to audio and video equipment (for example a TV) in a couple of different ways. Video output can be provided either through the composite output or through HDMI. XBMC is especially suitable for HDMI: using HDMI it can provide high definition video, a variety of high quality audio configurations, the ability to use the television's remote (in many TVs) and some control over turning the television on and selecting Raspberry Pi to view.

XBMC accepts video and audio in a very wide range of formats and can obtain material both from sources on the same computer or streamed across a home network (perhaps from a computer containing a music collection) or from across the Internet (from sources such as BBC iPlayer or YouTube). Because of its open design, however, it cannot play material that's protected by any form of Digital Rights Management (for example, virtually anything purchased through an online store). Devices capable of playing these kinds of content must have a means of proving that they're directly or indirectly authorized to use the content and to adhere to the supplier's access control policy. Players that can be altered by anyone, such as XBMC, aren't easily trusted by suppliers.

Unlike the standard X-based desktops available for Linux, XBMC requires hardware capable of providing 3D graphics. Luckily one of the interfaces that XBMC can use, OpenGL ES, is one that Raspberry Pi not only provides but provides really quite well, because it's supported through its graphics processors (discussed further in the annexe about multimedia libraries).

Getting XBMC
XBMC is available as a ready-built Raspberry-Pi SD card image that you can write to an SD card of your own.

SD card size
Naturally you'll need a blank SD card for this stage. These come in a variety of sizes. The software for Raspberry Pi will use less than 2GB, but because the SD card is also used to store your own files then the larger the card the better. A 4GB card should be considered a minimum size.

In addition to being able to play media files held on the SD card, XBMC can use those available from other computers in your house, or on an attached USB memory stick or hard drive (the connection of which we discuss in a separate section). However, if you want to use the SD card itself you'll probably need as large an SD card as possible for storing music and/or video files.

Downloading
There are several versions of XBMC available for Raspberry Pi and improvements are being made all the time. One, called OpenELEC (short for 'Open Embedded Linux Entertainment Centre'), is based on a minimal Linux distribution which is particularly fast to start up. The latest SD card images for this can be found at

    http://openelec.thestateofme.com/

Your first task is to download one of these on to a computer from which you can write it to an SD card. As with the reference SD card distribution, you'll need a computer with Web access to perform this download, an SD-card writer and some spare disk space.

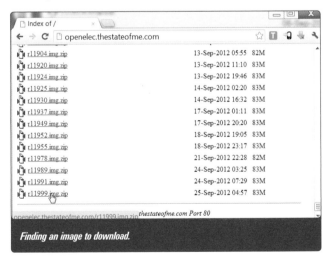

Finding an image to download.

Once downloaded unzip the file and write the resulting image file to your new SD card (see the section about creating an SD card for the first time).

Help
OpenElec have provided their own description of how to undertake this procedure at

    http://wiki.openelec.tv/index.php?title=
    Installing_OpenELEC_on_Raspberry_Pi

Expand SD card

This version of the SD card provides only a small area for storing media files. If you intend to use all of your SD card you'll need to extend the '/storage' file system and partition. The easiest way to do this is to edit the partitions on the SD card using a program called a 'partition editor' on another computer.

Using a TV remote control

Before we talk about the first time XBMC is run it's worth looking at your television remote control if you use an HDMI connection, because it may be able to perform a certain number of functions in place of a mouse or keyboard. If you're using Raspberry Pi in your sitting room this is normally by far the most convenient way to control XBMC.

This is a typical remote control. Although yours might be different from the one illustrated, the symbols used for the keys and the general layout of these sections are likely to be similar.

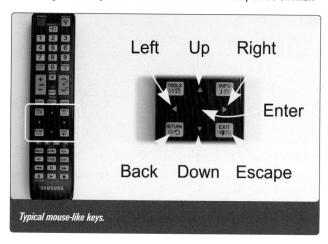

Typical mouse-like keys.

These keys are used by XBMC to navigate the menus. The up, down, left and right buttons behave as the arrow keys on a keyboard would and generally move something in the relevant direction. The central key is equivalent to the 'Enter' key and usually signals that XBMC should do something with what's been selected using the other keys. Back and Escape often return to the last menu page, although the Escape or Exit key may go further back.

In remotes that support recording and playback there's normally the following cluster of keys:

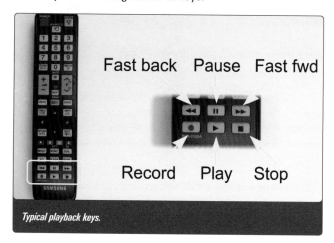

Typical playback keys.

These keys are used by XBMC to control the playback of music or video and have the usual interpretation.

First run

When OpenElec boots it displays an 'OpenElec' page almost immediately and displays the XBMC version number. It then takes several seconds to bring XBMC up properly, during which the screen is blank, before finally providing the main start-up screen which will probably not look exactly like this:

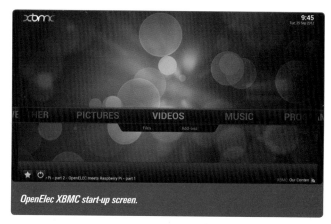

OpenElec XBMC start-up screen.

Setting the screen size

The reasons that what you see may not look like the above are that: the screen resolution starts off very low, so that not as much detail can be displayed; the aspect ratio (between the height and width of the screen) may not be the same as your display; and there might be too much overscan, so that the edges of the screen aren't visible. All of these problems can be solved easily.

To do this you need to select system settings using either your mouse or, possibly, your television remote control.

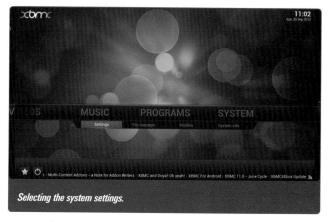

Selecting the system settings.

Where you can, select the 'Resolution' option to improve the quality of video. Most HDMI televisions will support the maximum resolution of 1920 x 1080, which you can select by operating the up and down arrowheads on the right-hand side of the line.

Once that's been done it's sensible to address any overscan problem you may have using the video calibration option under the 'video output' tab.

Video settings with HD resolution set.

This displays a small number of screens in sequence which invite you to adjust certain aspects of the display. The first sets the correct position for the top left of the display area for your screen. Although it may not be visible this screen has white lines at the very top and the very left of the display area. You should use the direction controls on either your keyboard (the arrow keys) or your TV remote to move the screen until both lines are visible, and you see something like this:

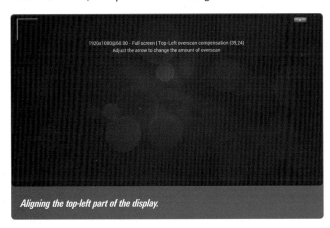

Aligning the top-left part of the display.

When you are satisfied press either the enter key on your keyboard or the key in the middle of the direction keys on your remote control to bring you to the next calibration screen.

Similarly the next screen should be adjusted until is looks like this:

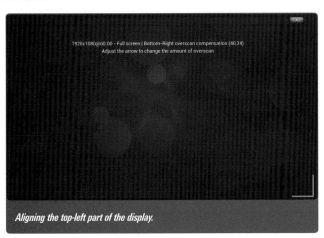

Aligning the top-left part of the display.

You may wish to perform further calibration as provided for in the following screens, but those were the most important issues to deal with.

Set your region

Initially you may find that the time isn't correct because XBMC isn't using the correct time zone information. This and other information about your region can be set using the 'International' tab of the 'Appearance' system setting.

Set the language, region and time zone of your choice here:

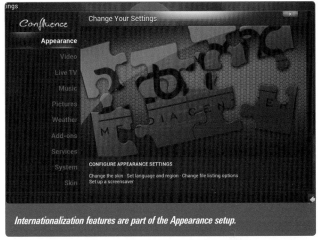

Internationalization features are part of the Appearance setup.

Language, region and time zone settings.

Join your home's network group

If you have one or more Windows computers in your home it's easy to access them from XBMC using the 'SMB' computer network protocol (which is discussed in a separate section). XBMC supports this without any additional setup but this protocol includes the notion of separate 'work groups' of computers. Discovering other computers in your home will be easier if you tell XBMC which work group name it should use.

Both Windows and XBMC will use the work group name 'WORKGROUP' if nothing else has been done. If you've set your own workgroup for your home computers you'll need to set this name.

This is set using the same system settings screen as before but using the 'services' screen under the 'SMB Client' tab.

Configuring XBMC's use of Windows shares.

Once you have changed the work group name you will need to reboot XBMC.

Changes to XBMC's workgroup require a reboot.

What XBMC can play

Combining both pictures and music, video is the most complex of the things that XBMC is asked to play. To understand its capabilities it's useful to know a little more about video files.

The type of a video file is usually indicated by the file name's 'extension', that is the three or four letters following its name. These letters are usually hidden from you when files are displayed in a folder on a desktop operating system but are evident in the icon used to represent the file and its file type property. There are a quite a number of types such as Microsoft's AVI, ASF, WMA and WMV; Adobe's FLV and F4V; the DVD Forum's VOB and IFO; Apple's MOV and QT; the Motion Picture Expert's Group MPG, MPEG, MP4 and PS, and so on. Most of these, however, determine only the kind of container that video and audio content is placed in, not the format of the video.

A container file includes internal information that indicates exactly which video (and audio) format it holds. XBMC is able to deal with all common container files.

To display a given video format XBMC needs a type of software called a codec (a shortened form of 'coder-decoder', although in this case we're only interested in decoding). The success or otherwise of XBMC being able to show a video is

very dependent on the complexity of the codec and whether it's provided as 'software', running on the ARM, or 'firmware', running on the VideoCore graphics processors. To the first approximation any video that uses a software codec is likely to play poorly, but most video that uses a firmware codec will play well.

To make things more complex virtually no codec is 'free' in the sense described in the introduction, instead they must be licensed, and someone has to pay a licence fee for it. Even the video files that you may enjoy on your desktop computer, set-top box or DVD player are decoded by codecs that have been paid for somehow.

The purchase price of a standard Raspberry Pi system includes the cost that the manufacturers had to pay to license two firmware codecs for the H.264 and MPEG4 formats. These are currently two of the most popular video formats. Possibly over half of the files you need to play will require an H.264 codec.

Two further codecs are available from the Raspberry Pi Foundation, one for VC-1 and one for MPEG2, for which a small fee is charged (so that the corresponding licence can be paid to the licence owners).

There are a small number of further video formats for which no firmware codec is available. This is a rough guide to the usefulness of each codec:

Codec	Use	Firmware
H.264	Blu-ray video, Internet streaming from YouTube, Vimeo, iTunes store, Freeview HD, and many file containers (eg Adobe's Flash container FLV).	Yes
MPEG4	Many container types including MP4, DIVX, Xvid and Apple's QuickTime containers (MOV, QT).	Yes
MPEG2	Some Blu-ray video, almost all DVD videos, Freeview (not HD).	Paid
VC-1	Most Microsoft Windows Media Video version 9 containers (MWV, ASF, MKV, AVI). Note that the same containers can also hold less popular WMV versions, which are in a completely different format.	Paid
Others	Not very common.	No

The success or otherwise of decoding a video also depends on other factors such as the speed with which the file can be delivered and the amount of effort the ARM has to put into obtaining the file.

Files held in Raspberry Pi's SD card or other directly attached storage are fastest, with files read over the home network being a little slower. Files read over the Internet are generally the slowest. This speed is normally relevant only for the higher resolution videos. For example, some 'HD' formats may not play well even over your home network if you've chosen a 10Mbps Ethernet connection or have WiFi with a poor signal.

The protocols used to read files over your home network might also have an effect. In previous sections we discussed the use of Windows shares to obtain files from other computers. This is noticeably more difficult for the ARM to process than the alternative Network File System (NFS) method of sharing files, so will reduce the other processing that the ARM may have available.

Finally any Digital Rights Management associated with the container file or codec (commonly enabled in Blu-ray, Windows codecs and most purchased video and audio) are likely to render a file unplayable. You may find this perverse, but many legal means to obtain copyright content will provide files that aren't playable.

Buying codecs

Codecs bought for your Raspberry Pi must be used with the board you own. If you have more than one Raspberry Pi you'll need to buy more than one codec, but the same purchase will cover the use of the codec on any number of systems that you use on it. (For example, if you buy an MPEG2 codec for XBMC you'll also be able to use the licence to allow its use under the reference Linux distribution.) To purchase a codec, visit the Raspberry Pi Foundation's website shop at

```
http://www.raspberrypi.com/
```

These are the steps involved:

- Find your serial number.
- Select codecs to purchase on the Web.
- Pay for them.
- Wait for email confirmation.
- Wait for email licences.
- Install licences on Raspberry Pi.

To buy any of these licences you must first provide the serial number of your Raspberry Pi single-board computer. The serial number is a 16-digit hexadecimal number. In XBMC this information can be found in the system information page about the 'Hardware':

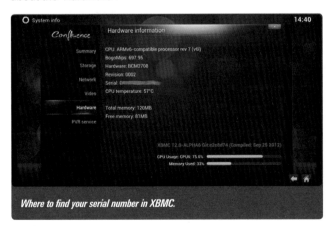

Where to find your serial number in XBMC.

At the website type this number in as you order each codec that you wish to buy.

Then proceed to the checkout, paying for the codec either with your PayPal account, if you have one, or with a credit card. During the process you must provide an email address. It's to this address that the licence is sent. After the purchase look for new emails from the Raspberry Pi Foundation. The first emails contain files describing how the licences are to be

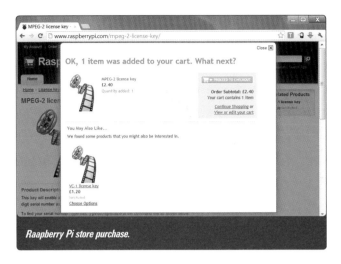

Raapberry Pi store purchase.

installed, but don't contain the licences themselves.

Eventually you should receive one email for each codec including an eight digit hexadecimal number, for example 0x12345678. In the annexe about configuration we discuss the purposes of the different files that can be found on the boot partition of a Raspberry Pi SD card. One of them, config.txt, provides various information to the program that boots Linux, including codec licences. To use the new codec you'll need to edit or create this file and ensure that it has a line saying

```
decode_MPG2=0x12345678
```

for the MEPG2 codec and/or

```
decode_WVC1=0x12345678
```

for the VC-1 codec.

On a new system this file may not exist so it'll have to be created. There are two ways to edit this file. The first is to remove the SD card from Raspberry Pi and plug it into another computer where the file can be edited. For example, on Windows use 'notepad': with the folder in 'Computer' that represents the SD card, right-click to obtain a menu of actions and select 'New' and then 'Text Document'. Type the name of the new file 'config.txt' and press the Enter key.

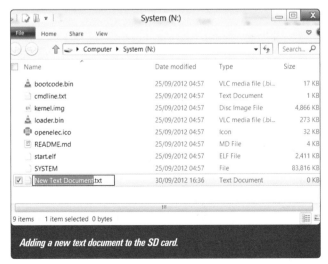

Adding a new text document to the SD card.

You can then enter the licences to this file in Notepad:

Finish using the menu File > Save, replace the SD card into Raspberry Pi and reboot it.

```
decode_MPG2=0x12345678
decode_WVC1=0x12345678
```

Entering the codec details into config.txt.

The other method is to log in to XBMC using SSH (which we describe how to do below) and edit the file with the Linux editor 'vi'. However, this is only recommended for those already familiar with this editor.

```
vi /flash/config.txt
```

Playing PC music

Many people already have a supply of music on another computer, managed perhaps by iTunes or Windows Media Player. For example, if you use iTunes on a Windows computer you may find that your music is stored in a subfolder of 'My Music' called 'iTunes\iTunes Music'. By sharing this folder you can make it available to XBMC.

To use music shared in this way you need to add a music source to XBMC, which you do by using the Music menu item, from where you can select 'Add Source'. This opens a dialogue that allows you to 'Browse'. Select 'Windows Shares' and find the machine and share with your music on it.

Finding a Windows share on your network.

Once you have done that give a meaningful name to this source and click on 'OK'.

Naming your new music source.

You should now find that a new entry has appeared under Music > Files with the name you gave it. Selecting it, it should be possible to find individual tracks or directories full of tracks to play.

Playing music 'full screen'.

Once playing it's possible to select an animation to go with it (add-ons are available) and to use the whole screen. Otherwise the track can be left playing in the background while some other activity is selected.

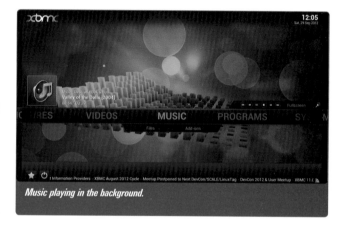

Music playing in the background.

Using Raspberry Pi as a DVD player

It's possible to use your Raspberry Pi as a stand-alone DVD player with the addition of just one extra piece of kit: a DVD (or Blu-ray) drive. Currently these are available for around £20. If you buy one for this purpose, though, remember that Raspberry Pi cannot provide a large current through its USB sockets. It's probably best to ensure that any drive you buy has an accompanying power supply (some do, even though working without one is a common selling point). Otherwise you should consider using the DVD drive through a powered USB hub (we discuss this in the section about setting Raspberry Pi up).

Once you've connected the DVD

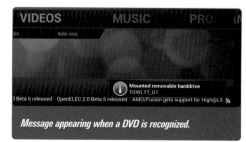

Message appearing when a DVD is recognized.

drive there's little involved in playing a DVD. When the drive is connected XBMC will provide an information message saying that the drive is available. The name provided will be

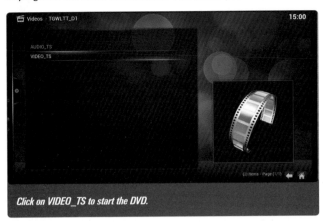

Click on VIDEO_TS to start the DVD.

one associated with the DVD in the DVD drive. The same name should appear in the list of video files. Clicking on it will usually provide two options (corresponding to directories on the DVD) 'AUDIO_TS' and 'VIDEO_TS'. Clicking on the latter should start the film.

Raspberry Pi playing a DVD.

Currently XBMC does not decode the DVD menus so the film will start immediately.

SMB access to XBMC

Most of the places where file locations are needed in XBMC have access to Windows shares on other computers. It's also the case that XBMC provides a number of shares itself, which are easily seen from other computers.

In order to make use of this feature only the IP address of Raspberry Pi is needed, which can be found from the systems

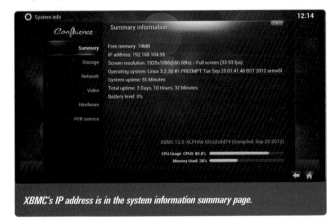

XBMC's IP address is in the system information summary page.

information display:
For example, if it were 192.168.104.95, typing this

```
\\192.168.104.95
```

into the location bar in Windows Explorer would give a folder display something like:

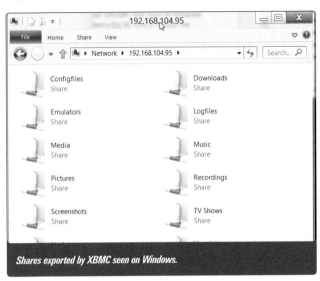

Shares exported by XBMC seen on Windows.

On Linux systems typing in 'smb://192.168.104.95/' will have the same effect:

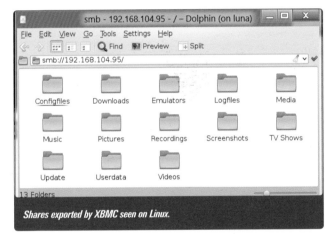

Shares exported by XBMC seen on Linux.

Among others there are shares for the directories that contain the media files that can be displayed as pictures, music and videos. This makes it quite simple to update what XBMC can show from another computer (since new material can be copied into these locations from your desktop computer).

Doing more with XBMC

XBMC.org provides documentation about its basic use, adding on new plugins, playing media files from all kinds of different sources including information and various more advanced topics at

```
http://wiki.xbmc.org/index.php
```

They even have a section for Raspberry Pi users.
XBMC forum members also publish a large number of ad-hoc guides which you can find at

```
http://forum.xbmc.org/forumdisplay.php?fid=110
```

07.
Annexes

ANNEXES
Configuration

If a file called 'config.txt' exists on the boot partition (/boot from the reference Raspberry Pi distribution) it'll be used to configure various properties of Raspberry Pi before the main ARM loader begins execution. Each line of the file should be one of the following:

Command	Description
`<empty>`	Does nothing.
`#<comment>`	Ignores the comment.
`<property>=<integer>`	Sets the property to the number given.
`<property> <integer>`	Also sets the property to the number given.
`<configuration file command>`	One of the 'bootcfg.txt' commands described below.

Many of the properties that can be used are described in the following sections. Note, however, that the precise set of properties can change from release to release of Raspberry Pi's firmware.

Overclocking properties

Like all chips, the BCM2835 is specified quite conservatively in order to ensure that the greatest percentage meet the published specification. The result is that your Raspberry Pi might well be able to operate at a slightly faster speed (although it might not). It's quite common to find that it'll operate at 900MHz. The following properties allow you to change the voltage or the speed of the clock delivered to different sections of the BCM 2835 in Raspberry Pi.

If your chip isn't one of those that can tolerate higher clock speeds it'll behave erratically (and you should readjust your settings). Some higher speeds can be maintained if you also deliver a higher voltage. Unfortunately, although small changes aren't very likely to make a difference, it isn't entirely safe to deliver higher voltages to the chip. You could do permanent damage to it. For this reason the settings that allow you to increase voltages will blow a One Time Programmable (OTP) bit when they're used, which you cannot repair. If you were to return a system as faulty with this bit set it wouldn't be covered by your warranty.

One final warning is that the lifetime of your Raspberry Pi is shortened by running it at higher speeds and voltages, although its planned lifetime is quite long (tens of years) in the first place.

The 'force_turbo' option (which isn't set by default) determines whether the system manages its own clock frequencies or not (usually, when it isn't set it does).

Property	Description
arm_freq	Force the frequency of the ARM (in MHz) defaulting to 700. When force_turbo isn't set this is the maximum frequency that will be used, not the constant frequency.
gpu_freq	Force the frequency of the GPU (in MHz). Equivalent to setting core_freq, h264_freq, isp_freq and v3d_freq separately.
core_freq	Force the system frequency (in MHz) defaulting to 250. When force_turbo isn't set this is the maximum frequency that will be used, not the constant frequency.
h264_freq	Force the frequency of the H.264 unit (in MHz) defaulting to 250. Unused unless 'turbo_mode' is set.
isp_freq	Force the frequency of the ISP unit (in MHz) defaulting to 250. Unused unless 'turbo_mode' is set.
v3d_freq	Force the frequency of the V3D unit (in MHz) defaulting to 250. Unused unless 'turbo_mode' is set.
avoid_pwm_pll	Allow the ARM, SDRAM and GPU to use frequencies that are unrelated to the GPU (the core, H.264, v3d and ISP components are always obliged to have tightly coupled clock frequencies).
sdram_freq	Force the frequency of the main SDRAM memory (in MHz) defaulting to 400MHz. When force_turbo isn't set this is the maximum frequency that will be used, not the constant frequency.
never_over_voltage	Ignore any other setting that might request artificially high voltages. This blows an OTP bit and is permanent.
over_voltage	Set higher system voltage (up to 31 in units of 0.025V). Note that use of this option will be registered by blowing an OTP bit, indicating that the board warranty is voided. Values of 8 or greater require 'force_turbo' to be set.
over_voltage_sdram	Set higher voltages on controller, I/O and PHY SDRAM (effectively sets over_voltage_sdram_c, over_voltage_sdram_i and over_voltage_sdram_p). Note that use of this option will be registered by blowing an OTP bit, indicating that the board warranty is voided.
over_voltage_sdram_c	Set a higher voltage on the SDRAM controller only. Values can range between -16 to 8 in units of 0.025V, with zero being 1.2V. Only effective when 'force_turbo' is set. Note that use of this option will be registered by blowing an OTP bit, indicating that the board warranty is voided.
over_voltage_sdram_i	Set a higher voltage on I/O SDRAM only. Values can range between -16 to 8 in units of 0.025V with zero being 1.2V. Only effective when 'force_turbo' is set. Note that use of this option will be registered by blowing an OTP bit, indicating that the board warranty is voided.
over_voltage_sdram_p	Set a higher voltage on PHY SDRAM only. Values can range between -16 to 8 in units of 0.025V with zero being 1.2V. Only effective when 'force_turbo' is set. Note that use of this option will be registered by blowing an OTP bit, indicating that the board warranty is voided.

Property	Description
force_turbo	Enables the h264_freq, v3d_freq, isp_freq overclocking, disables the settings ending in '_min' below and disables the driver that changes the system frequencies dynamically.
initial_turbo	Enters the turbo mode (described by force_turbo) from boot for the given number of seconds.
arm_freq_min	The minimum value of arm_freq used when adjusting system frequencies dynamically (that is when 'force_turbo' isn't set).
core_freq_min	The minimum value of core_freq used when adjusting system frequencies dynamically (that is when 'force_turbo' isn't set).
sdram_freq_min	The minimum value of sdram_freq used when adjusting system frequencies dynamically (that is when 'force_turbo' isn't set).
over_voltage_min	The minimum value of over_voltage used when adjusting system frequencies dynamically (that is when 'force_turbo' isn't set).
temp_limit	The number of degrees Celsius at which reduced system frequencies and voltages are automatically applied to cool the system. Setting this value greater than the default of 85 destroys the warranty OTP bit.
current_limit_override	Set password (0x5A000020) to allow a current limit override when SDRAM over voltage is set. Setting this option will also destroy the warranty OTP bit.

HDMI and display configuration properties

Raspberry Pi will normally detect whether you have an HDMI display plugged in and, if found, use that in preference to composite video output. If the HDMI display is found it's normally queried to discover which resolutions it supports, and one that's acceptable to Raspberry Pi is then used.

Some HDMI displays are less than perfect and don't always give expected results, so a number of these properties deal with overriding the information that the display should provide.

Property	Description
sdtv_mode	Sets the standard to be used for composite video output: 0 - Normal NTSC, 1 - Japanese version of NTSC - no pedestal, 2 - Normal PAL, 3 - Brazilian version of PAL (PAL/M) - 525/60 rather than 625/50, different sub-carrier.
sdtv_aspect	Set the ratio between the screen width and height for composite video output: 1 - SDVT aspect ratio 4:3, 2 - SDVT aspect ratio14:9, 3 - SDVT aspect ratio16:9.
hdmi_safe	Provide a safe setting for a maximum compatibility with HDMI. Equivalent to hdmi_force_hotplug (set), hdmi_ignore_edid, config_hdmi_boost (4), hdmi_group (set), hdmi_mode (1) and disable_overscan (unset).
hdmi_mode	Sets the screen resolution. The available HDMI modes depend on hdmi_group. The values are listed separately below.
hdmi_group	By default the HDMI group reported by the EDID will be used, otherwise: 1 – CEA HDMI modes. 2 – DMT HDMI modes.

Property	Description
hdmi_drive	1 - DVI mode (no sound). 2 - HDMI mode (sound will be sent if possible).
hdmi_force_hotplug	Ignore hotplug pin on HDMI connector and assume it's always asserted. (This means that composite video won't be selected when the HDMI connector doesn't deliver this signal.)
hdmi_pixel_encoding	0 - RGB_LIMITED. 1 - RGB_FULL. 2 - YCbCr444_LIMITED. 3 - YCbCr444_FULL.
hdmi_pixel_clock_type	0=PAL. 1=NTSC (1000/1001).
edid_content_type	Force EDID content type byte: bit0=text. bit1=photo. bit2=cinema. bit3=game. bit5=video_present. bit6=interlaced_latency. bit7=progressive_latency.
disable_overscan	Don't reduce the area of the display natively provided (ie assume the device doesn't overscan). Ignore overscan_leff, overscan_right, overscan_top and overscan_bottom.
overscan_left	Left-hand margin from natively displayed (overscanned) image left-hand.
overscan_right	Right-hand margin from natively displayed (overscanned) image right-hand.
overscan_top	Top depth from natively displayed (overscanned) image top.
overscan_bottom	Bottom rise from natively displayed (overscanned) image bottom.
framebuffer_width	Force the initial console framebuffer width, rather than default of display's size minus overscan.
framebuffer_height	Force the initial console framebuffer height, rather than default of display's size minus overscan.
framebuffer_depth	Force the initial console framebuffer depth in bits per pixel. The default is 16. 8 bits per pixel is valid, but the default RGB palette makes an unreadable screen. 24 bits per pixel looks better but may be unstable. 32 bits per pixel will need framebuffer_ignore_alpha=1 but may not show accurate colours.
hdmi_clock_change_limit	Limit HDMI clock change limit with syncing display to video frame-rate. In ppm.
enable_hdmi_status	Enable the display of the HDMI status on the screen.
config_hdmi_boost	Set a boost to the HDMI signal (increased to at least 2).
config_hdmi_preemphasis	Set a pre-emphasis (between 1 and 7) to the HDMI signal (also increased by an HDMI boost). Try 4 if you seem to have issues with HDMI interference.
spread_spectrum_enable	Enable spread spectrum clocking.
frequency_deviation_ppm	Deviation from the central frequency in parts per million, used when spread_spectrum_enable is set.
modulation_frequency_hz	Frequency over which the deviating frequency is cycled when spread_spectrum_enable is set.
hdmi_pixel_freq_limit	Maximum pixel clock for HDMI.

Property	Description
hdmi_clock_spreading	HDMI clock frequency to vary during clock spreading in parts per million.
hvs_swap_red_blue	Swap red and blue at output of HVS.
hdmi_ignore_edid	If set to magic value 0xa5000080, the EDID is ignored, and any hdmi_group/hdmi_mode can be driven. (Sometimes required by Chinese displays.)
hdmi_force_edid_audio	Pretend all HDMI audio formats (eg DTS/AC3) are supported even if not reported in EDID.
hdmi_edid_file	Don't read EDID from device, but load it from file 'edid.dat' in boot directory (on SD card).
framebuffer_ignore_alpha	Ignore alpha in framebuffer. (This is particularly useful when using a 32-bit framebuffer depth.)
framebuffer_align	Force alignment of framebuffer allocation to the number of bytes given to this property.
disable_splash	Set non-zero to disable rainbow splash screen during boot.
display_rotate	0 – no rotation. 1 – 90°. 2 – 180°. 3 – 270°.

HDMI modes

This is a table of the values that can be used for the HDMI mode hdmi_mode. The interpretation of the value depends on which hdmi_group has been selected.

In the table 'H' indicates a 16:9 aspect ratio in a mode that would otherwise have a 4:3 aspect ratio. '2x' indicates modes where each pixel is sent twice (doubling the clock rate) and '4x' modes where each pixel is sent four times.

Mode	hdmi_group CEA	hdmi_group DMT
1	VGA	640x350 85Hz
2	480p 60Hz	640x400 85Hz
3	480p 60Hz H	720x400 85Hz
4	720p 60Hz	640x480 60Hz
5	1080i 60Hz	640x480 72Hz
6	480i 60Hz	640x480 75Hz
7	480i 60Hz H	640x480 85Hz
8	240p 60Hz	800x600 56Hz
9	240p 60Hz H	800x600 60Hz
10	480i 60Hz 4x	800x600 72Hz
11	480i 60Hz 4x H	800x600 75Hz
12	240p 60Hz 4x	800x600 85Hz
13	240p 60Hz 4x H	800x600 120Hz
14	480p 60Hz 2x	848x480 60Hz
15	480p 60Hz 2x H	1024x768 43Hz DO NOT USE
16	1080p 60Hz	1024x768 60Hz
17	576p 50Hz	1024x768 70Hz
18	576p 50Hz H	1024x768 75Hz
19	720p 50Hz	1024x768 85Hz
20	1080i 50Hz	1024x768 120Hz
21	576i 50Hz	1152x864 75Hz
22	576i 50Hz H	1280x768 reduced blanking
23	288p 50Hz	1280x768 60Hz
24	288p 50Hz H	1280x768 75Hz
25	576i 50Hz 4x	1280x768 85Hz
26	576i 50Hz 4x H	1280x768 120Hz reduced blanking
27	288p 50Hz 4x	1280x800 reduced blanking
28	288p 50Hz 4x H	1280x800 60Hz
29	576p 50Hz 2x	1280x800 75Hz
30	576p 50Hz 2x H	1280x800 85Hz
31	1080p 50Hz	1280x800 120Hz reduced blanking
32	1080p 24Hz	1280x960 60Hz
33	1080p 25Hz	1280x960 85Hz
34	1080p 30Hz	1280x960 120Hz reduced blanking
35	480p 60Hz 4x	1280x1024 60Hz
36	480p 60Hz 4xH	1280x1024 75Hz
37	576p 50Hz 4x	1280x1024 85Hz
38	576p 50Hz 4x H	1280x1024 120Hz reduced blanking
39	1080i 50Hz reduced blanking	1360x768 60Hz
40	1080i 100Hz	1360x768 120Hz reduced blanking
41	720p 100Hz	1400x1050 reduced blanking
42	576p 100Hz	1400x1050 60Hz
43	576p 100Hz H	1400x1050 75Hz
44	576i 100Hz	1400x1050 85Hz
45	576i 100Hz H	1400x1050 120Hz reduced blanking
46	1080i 120Hz	1440x900 reduced blanking
47	720p 120Hz	1440x900 60Hz
48	480p 120Hz	1440x900 75Hz
49	480p 120Hz H	1440x900 85Hz
50	480i 120Hz	1440x900 120Hz reduced blanking
51	480i 120Hz H	1600x1200 60Hz
52	576p 200Hz	1600x1200 65Hz
53	576p 200Hz H	1600x1200 70Hz
54	576i 200Hz	1600x1200 75Hz
55	576i 200Hz H	1600x1200 85Hz
56	480p 240Hz	1600x1200 120Hz reduced blanking
57	480p 240Hz H	1680x1050 reduced blanking
58	480i 240Hz	1680x1050 60Hz
59	480i 240Hz H	1680x1050 75Hz
60		1680x1050 85Hz
61		1680x1050 120Hz reduced blanking
62		1792x1344 60Hz
63		1792x1344 75Hz
64		1792x1344 120Hz reduced blanking
65		1856x1392 60Hz
66		1856x1392 75Hz
67		1856x1392 120Hz reduced blanking
68		1920x1200 reduced blanking
69		1920x1200 60Hz
70		1920x1200 75Hz
71		1920x1200 85Hz
72		1920x1200 120Hz reduced blanking
73		1920x1440 60Hz
74		1920x1440 75Hz
75		1920x1440 120Hz reduced blanking
76		2560x1600 reduced blanking
77		2560x1600 60Hz
78		2560x1600 75Hz
79		2560x1600 85Hz
80		2560x1600 120Hz reduced blanking
81		1366x768 60Hz
82		1080p 60Hz
83		1600x900 reduced blanking
84		2048x1152 reduced blanking
85		720p 60Hz
86		1366x768 reduced blanking

HDMI setup example

An example 'config.txt' file might contain

```
sdtv_mode=2
sdtv_aspect=3
overscan_left=28
overscan_right=28
overscan_top=16
overscan_bottom=16
# arm_freq=600
```

TV setup

Your HDMI monitor may support only a limited set of formats. There's a small utility 'tvservice' provided on the reference Raspberry Pi SD card which enables the supported formats to be found (among other things).

- Set the output format to VGA 60Hz (hdmi_group=1 hdmi_mode=1) and boot up the Raspberry Pi.
- Enter the following command to give a list of CEA supported modes:

```
/opt/vc/bin/tvservice -m CEA
```

- Enter the following command to give a list of DMT supported modes:

```
/opt/vc/bin/tvservice -m DMT
```

- Use this command to show your current state:

```
/opt/vc/bin/tvservice -s
```

These commands can be used to provide more detailed information from your monitor:

```
/opt/vc/bin/tvservice -d edid.dat
/opt/vc/bin/edidparser edid.dat
```

The file edit.dat is what you'd need if you wanted to describe your HDMI setup to a third party (for example, on the Raspberry Pi forums website).

GPIO configuration properties

Property	Description
gpio_pads0	GPIO pad drive strength assigned to PM_PADS0.
gpio_pads2	GPIO pad drive strength assigned to PM_PADS2.
gpio_pads3	GPIO pad drive strength assigned to PM_PADS3.
gpio_pads4	GPIO pad drive strength assigned to PM_PADS4.
gpio_pads5	GPIO pad drive strength assigned to PM_PADS5.

UART configuration properties

These properties allow some control over the way that the UART is set up before the kernel is loaded.

Property	Description
init_uart_baud	UART baud rate in Hz (115200 by default).
init_uart_clock	Clock used for the UART controller in Hz (3,000,000 by default).

SD controller configuration properties

This configuration property provides some control over the part of the Raspberry Pi chip that deals with the SD memory card socket.

Property	Description
init_emmc_clock	Clock used for the Arasan SD controller in Hz (50,000,000 by default).

ARM loader configuration properties

The ARM Loader runs on VideoCore and loads the Linux kernel into the ARM.

Property	Description
boot_delay	A delay inserted after the loader is called before the kernel is loaded from the boot media (seconds).
boot_delay_ms	An additional part of the boot delay specified in milliseconds.
disable_pvt	Use an alternative PVT calibration schedule.
kernel_old	Use old-style kernel loading (ie don't prepend 'kernel.img' with 'start32.bin').
logging_force	Force the overall and message VCOS logging levels to the bottom nibble and the adjacent one of the value set respectively. (This logging is available through the VideoCore debugger.)
disable_commandline_tags	Don't automatically set command-line information in the 'ATAGS' memory area at address 0x100 (the kernel will expect this information to be provided manually before the kernel image).
ramfsaddr	Unused.
test_mode	Displays a test card on composite and plays analogue audio for <n> seconds before continuing with boot.
disable_l2cache	Whether the BCM 2835 second level cache is to be used for normal memory accesses from the ARM (must correspond to an analogous setting built into a new kernel). Unset by default.
device_tree_address	The address in memory where the contents of the device tree file (if provided) is placed (0x100 by default).
kernel_address	The address in memory where the contents of the kernel image file are placed (0x8000 by default).

Codec licensing

We discuss codecs, which video files they help display and how they can be licensed in the section about the XBMC media player. Codec licences are provided by means of a hexadecimal number that's linked to your Raspberry Pi's serial number. Codecs can be used only when the correct licence number is supplied to these option names.

Property	Description
decode_MPG2	A list of up to eight comma-separated numbers, one of which was provided when the MPEG-2 codec was purchased.
decode_WVC1	A list of up to eight comma-separated numbers, one of which was provided when the VC1 codec was purchased.
decode_DDP	A list of up to eight comma-separated numbers, one of which was provided when the DDP (Dolby Digital Plus), or AC3, codec was purchased.
decode_DTS	A list of up to eight comma-separated numbers, one of which was provided when the DTS codec was purchased.

The option values allow a list of numbers so that one SD card can be shared by a small group of (up to eight) Raspberry Pis.

Safe mode

The configuration options are easily diverse enough to create a set of files in the boot partition that'll prevent booting on every occasion. (Overclocking too aggressively, for example.)

If you have another computer with an SD card reader it's possible to recover from these situations simply by plugging the SD card into that computer and changing the updated files.

If you have no other computer however Raspberry Pi's "safe mode" has been provided just for you. To enter safe mode you must join together two specific pins (numbered 3 and 4)

on the Expansion Header. Special connectors are available that could be used, but wrapping a bare wire around both terminals will also work. Be very careful

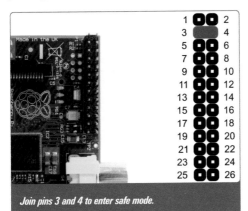

Join pins 3 and 4 to enter safe mode.

to use pins 3 and 4. Other pairs of pins may cause permanent damage to Raspberry Pi if they are joined together. It is unlikely to survive shorting together pins 1 and 2 for example (which supply two different voltages).

Safe mode ignores the files config.txt and cmdline.txt so that Raspberry Pi boots with all defaults in place. If there's a kernel on the boot partition called 'kernel_emergency.img', that kernel is used instead of the normal one (the normal kernel is used otherwise).

The important thing about the emergency kernel is that it doesn't need to use the SD card in order to run. It won't be prevented from booting, for example, if the root file system on the SD card has become corrupt. One consequence of this is that its root file system (which is held in memory) is very

small. It doesn't incorporate all the commands that you may be used to.

The commands that it does have are ones that are particularly useful for returning the configuration files to normal or for repairing the SD card's root partition.

Because you can place the SD card in another Windows (or, even better, Linux) computer, safe mode isn't your last chance to save your SD card before you junk all your files and write a new SD card image.

Other options

You can probably find an up-to-date list of the configuration options supported by the latest Raspberry Pi distribution on the Web at

    http://elinux.org/RPi_config.txt

The information there also includes some overclocking settings that have been successful.

The boot file system

The boot partition is home to the files that are needed to boot the Linux kernel. Conveniently, it's formatted in a way that will be understood by almost all operating systems. If you have a Windows computer with an SD card reader, for example, you should be able to change these files simply by inserting the SD card into the reader. This isn't true of the files in the second, root partition, where Linux files are kept.

This is a summary of the files that you may find there:

File name	Purpose
Bootcode.bin	The primary stage boot loader (boots Start.bin).
Start.elf	The ARM loader that boots Kernel.img and provides GPU services.
Start_cd.elf	A cut-down ARM loader that provides fewer GPU services.
Kernel.img	The normal ARM Linux kernel.
Kernel_cd.img	The cut-down kernel used when the GPU is given a small memory allocation.
Kernel_emergency.img	A kernel with a small built-in root file system that can be used in place of Kernel.img and which is used in 'safe mode'.
Fixup.dat	Information used when setting the GPU memory size in Kernel.img.
Fixup_cd.dat	Information used when setting the GPU memory size in Kernel._cdimg.
Cmdline.txt	Linux kernel command-line text.
Config.txt	ARM loader configuration properties.
Bootcfg.txt	File names configuration file containing file names configuration commands.
<file named in Bootcfg file (include)>	More file names configuration commands.
<file named in Bootcfg file (device_tree)>	Linux kernel 'device tree' description of the Raspberry Pi hardware.
<file named in Bootcfg file (initramfs)>	Image to use for an initial in-RAM file system by the Linux kernel.
Edid.dat	File used when hdmi_edid_file property is set.

ANNEXES
Multimedia Libraries

Raspberry Pi contains three separate processors. Only one (the ARM) is used by Linux. The other two are VideoCore processors, as we describe in the introduction. The job of these extra processors is varied but includes the provision of various audio and video services to Linux. The features they supply are available through a small number of standard libraries that can replace those provided by an ARM Linux distributor.

Any library can be provided in two formats. One, a static library, is used at the time a program is compiled and the relevant parts of it become part of that program's executable. The other, a dynamic library, is kept in the file system and used only at the last moment when the executable is run. Both forms have their own advantages. Most of the multimedia libraries are provided as dynamic libraries (which end with the suffix '.so', for shared object) and a few additionally have static versions (which end in '.a', for archive).

The libraries

The distribution supplies a set of open source but Raspberry Pi-specific libraries that give access to the GPU acceleration features. The libraries available are OpenGL, OpenVG, EGL and OpenMAX IL.

■ **OpenGL ES 2.0 (and OpenGL ES 1.1):** OpenGL is a 3D library, very commonly used on desktops and embedded systems. It's defined by the Khronos Group.
■ **OpenVG 1.1:** OpenVG is a 2D vector drawing library, also commonly used on desktops and embedded systems. Again, defined by the Khronos Group.
■ **EGL 1.4:** EGL is an interface between Khronos rendering APIs such as OpenGL ES or OpenVG and the underlying native platform window system.
■ **OpenMAX IL 1.1:** OpenMAX supplies a set of APIs that provides abstractions for routines used during audio, video and still images processing. OpenMAX defines three layers; this is the IL layer, which provides an interface between media framework and a set of multimedia components (such as codecs).

The first of these adhere to the standard Linux library APIs, so should be a straightforward swap-in for applications that already use those libraries. OpenMAX IL doesn't have a standard API at this stage, so is a custom implementation.

Codecs and open source components

The VideoCore Graphics Processing Unit (GPU) can hardware-decode H.264, MPEG1/2/4, VC1, AVS, MJPG at 1080p30. It can software-accelerate VP6, VP7, VP8, RV, Theora, WMV9 at DVD resolutions. The availability of some of the codecs is constrained by licensing but others are licence-free (for example, VP8, MJPG and Theora). We discuss the purchase and availability of codecs in the section about XBMC.

There are a few kernel drivers for the system provided in the kernel, which have been given a GNU Public Licence and hence are open source software. One of these drivers is the interface from the user space libraries to the GPU (the 'vchiq' driver). The user-side libraries use this driver to communicate with the GPU and tell it what to do. Being specific to the GPU these drivers don't provide general-purpose code that could be used on other types of GPU.

Many of the accelerated codecs are provided as GPU code that runs on the VideoCore processor.

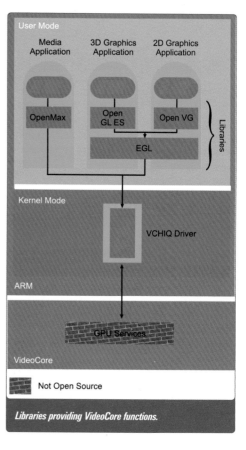

Libraries providing VideoCore functions.

File system location and installation

In general the name and location of each of these libraries, and of the relevant headers, is a property of the specific ARM Linux distribution. The reference Raspberry Pi distribution already includes these libraries so you don't need to install them yourself. In other distributions the libraries could be placed in the same place, being completely contained inside the directory

```
/opt/vc
```

This includes the subdirectories

■ /opt/vc/lib
■ /opt/vc/include
■ /opt/vc/bin

The following can be used to incorporate use of these libraries in preference to any similarly named libraries in a different ARM Linux distribution:

```
echo '/opt/vc/lib' | sudo tee /etc/ld.so.
  conf.d/_vmcs.conf
sudo ldconfig
```

Beware, however, that specific distributions will provide their own names for these libraries and their own library version numbering, and that existing applications that use the libraries will depend on the use of those names and version numbers. It may be more appropriate to inspect the relevant libraries using

```
ldconfig -p
```

and physically replace them with symbolic links into /opt/vc/lib.

Distributions will often provide packages for their own non-accelerated versions of these libraries. Secondary 'dev' packages provide standard Khronos headers in /usr/include. For example, Ubuntu 10.10 has a package 'libgles2-mesa' which provides /usr/lib/libGLESv2.so.2 for OPEN GL ES 2, and a separate package 'libgles2-mesa-dev' providing the headers

- /usr/include/GLES2/gl2.h
- /usr/include/GLES2/gl2ext.h
- /usr/include/GLES2/gl2platform.h

If your distribution doesn't provide a package with the relevant headers you can use the ones in /opt/vc/include.

OpenMAX IL 1.1

The Khronos Open Media Acceleration framework 'OpenMAX' provides APIs for audio, video and still images.

It provides interfaces at three layers of abstraction: Application, Integration and Development. Raspberry Pi's VideoCore processors provide support for the Integration layer ('OpenMAX IL').

This layer is responsible for creating a directed graph of streaming components (such as sources, sinks, codecs, filters, splitters, mixers, etc) determining the flow of information and order of processing used to achieve a desired multi-media task.

Khronos provides several significant resources to help understand this interface at

```
http://www.khronos.org/registry/omxil/
```

The current specification is held at

```
http://www.khronos.org/registry/omxil/specs/
OpenMAX_IL_1_1_2_Specification.pdf
```

The library providing accelerated access to these interfaces is available in /opt/vc/lib/libopenmaxil.so, and the standard headers can be found in /opt/vc/include. They include the headers

- IL/OMX_Audio.h
- IL/OMX_Broadcom.h
- IL/OMX_Component.h
- IL/OMX_ContentPipe.h
- IL/OMX_Core.h
- IL/OMX_ILCS.h
- IL/OMX_Image.h
- IL/OMX_Index.h
- IL/OMX_IVCommon.h
- IL/OMX_Other.h
- IL/OMX_Types.h
- IL/OMX_Video.h

Broadcom-specific extensions are provided by a plugins

mechanism supported by shared object libraries that would be placed in /opt/vc/lib/plugins/, but which aren't present in Raspberry Pi.

OpenGL ES 1.1 and 2.0

OpenGL ES is a cut-down version of the 'OpenGL' 3D graphics API that's been tailored to embedded systems.

Two versions of this interface are provided, one for OpenGL ES 1.1 and one for OpenGL ES 2.0. The specification for OpenGL ES 1.1 is held at

```
http://www.khronos.org/registry/gles/
specs/1.1/es_full_spec_1.1.12.pdf
```

The specification for OpenGL ES 2.0 is held at

```
http://www.khronos.org/registry/gles/
specs/2.0/es_full_spec_2.0.25.pdf
```

This API also has manual pages available online at

```
http://www.khronos.org/opengles/sdk/
docs/man/
```

Khronos also provides many significant resources regarding this interface at

```
http://www.khronos.org/registry/gles/
```

The library providing accelerated access to these interfaces is available in /opt/vc/lib/libGLESv2.so with a static version of the library at /opt/vc/lib/libGLESv2_static.a. The standard headers can be found in /opt/vc/include, and include

- GLES/gl.h
- GLES/glext.h
- GLES/glplatform.h

for OpenGL ES 1.1, and

- GLES2/gl2.h
- GLES2/gl2ext.h
- GLES2/gl2platform.h

for OpenGL ES 2.0.

OpenVG 1.1

Khronos provides many resources regarding this interface at

```
http://www.khronos.org/registry/vg/
```

An introductory presentation about its use can be found on the Khronos website:

```
http://www.khronos.org/developers/library/
siggraph2006/Khronos_Tech_Talks/HUONE_
OpenVG-How-to-Program-advanced-2D-vector-
graphics.ppt
```

The current specification is held at

```
http://www.khronos.org/registry/vg/specs/
openvg-1.1.pdf
```

The library providing accelerated access to these interfaces is available in /opt/vc/lib/libOpenVG.so. The standard headers can be found in /opt/vc/include, and include

- VG/vgu.h
- VG/vgplatform.h
- VG/vgext.h
- VG/openvg.h

EGL 1.4

EGL provides mechanisms for creating and managing the rendering surfaces on to which client APIs, such as OpenGL ES and OpenVG, can draw. It creates 'graphics contexts' for client APIs, and synchronizes drawing from both client APIs and native VideoCore rendering APIs. (Khronos provides an overview at http://www.khronos.org/egl/.)

Khronos provides resources regarding this interface at

```
http://www.khronos.org/registry/egl/
```

The current specification is held at

```
http://www.khronos.org/registry/egl/specs/
  eglspec.1.4.20101006.pdf
```

The library providing accelerated access to these interfaces is available in /opt/vc/lib/libEGL.so with a static version at /opt/vc/lib/libEGL_static.a. The standard headers can be found in /opt/vc/include, and include

- EGL/egl.h
- EGL/eglext.h
- EGL/eglplatform.h
- EGL/eglext_brcm.h

ANNEXES
The Demo C Programs

The reference SD card includes a demonstration of the multimedia capabilities of the Raspberry Pi. It's written in C and can be found in the reference Raspberry Pi image in this directory:

```
/opt/vc/src/hello_pi
```

Here you'll find at least the following directories:

Name	Purpose
hello_world	The archetypal simple C program.
hello_audio	Making a sound from scratch using the OpenMAX IL interface.
hello_video	Displaying a video file using OpenMAX IL.
hello_triangle	Displaying a rotating cube of six textures using Open GL ES 2 and EGL interfaces.
hello_triangle2	Demonstrates the use of Open GL ES 2 shader programs to display a fractal.
hello_dispmanx	Demonstrates the provision of hardware overlays.
hello_encode	Shows how to use the VideoCore to perform video encoding on your behalf.
hello_tiger	Displays a complicated vector-graphics image using the Open VG interface.
hello_font	Demonstrates how a subtitle can be rendered using a True Type font.

These are each C programs of varying complexity that show how different multimedia features provided by Raspberry Pi's VideoCore processors can be used.

Preparation

These demonstrations are available on the reference Raspberry Pi SD card (which you probably have). To use them you'll need to have a command line shell available (which we described in the section about Linux). If you're using the graphical desktop (LXDE) you should run the terminal emulator 'LXterminal' to get one.

Hello World

This code can be found in

```
/opt/vc/src/hello_pi/hello_world
```

so to start you should select this as your working directory using

```
cd /opt/vc/src/hello_pi/hello_world
```

The C program we're interested in is called 'hello_world.c' and you can display it using

```
less hello_world.c
```

You'll see that by far the majority of the code is a comment describing a 'BSD'-style licence, where Broadcom claims the copyright but allows you to do just about anything you like with it as long as you keep the comment and don't expect it to work in any way whatsoever. This is pretty typical of all source files in the open-source world, and you might consider putting the same kind of comment at the front of your own programs: but it does look a little odd in front of a four-line program.

The program itself includes a standard C header file 'stdio.h' which includes the definition of several functions that read and write data including 'printf', which is used in the function that's always called first in C 'main'. The effect of the program is simply to print 'Hello World!' to the screen. The '\n' at the end of the string argument to printf represents the end-of-line character in C, and when printed subsequent characters will appear on the next line.

The other file in this directory is Makefile, which provides the instructions for creating an application from the C code. In this case the Makefile just uses the instructions from the 'Makefile.include' file in the parent directory. (This allows the same 'make' instructions to be used for all of the demo projects.) The main instructions here describe how to make an 'object' file from a C file, and then an application from one or more object files and some libraries. An 'object' file is created by a compiler and contains all the machine code corresponding to the lines of C in the program being compiled. It cannot be run as a program, though, until it's been combined with other object code that provides machine code for any external functions that it's used (such as 'printf' in this case). This combination process is called 'linking'.

To obey the Makefile simply type the command

```
make
```

You should see a surprising amount of output, like

```
cc -DSTANDALONE -D__STDC_CONSTANT_MACROS -D__STDC_LIMIT_MACROS
-DTARGET_POSIX -D_LINUX -fPIC -DPIC -D_REENTRANT -D_LARGEFILE64_
SOURCE -D_FILE_OFFSET_BITS=64 -U_FORTIFY_SOURCE -Wall -g
-DHAVE_LIBOPENMAX=2 -DOMX -DOMX_SKIP64BIT -ftree-vectorize -pipe
-DUSE_EXTERNAL_OMX -DHAVE_LIBBCM_HOST -DUSE_EXTERNAL_LIBBCM_HOST
-DUSE_VCHIQ_ARM -Wno-psabi -I/opt/vc/include/ -I/opt/vc/include/
interface/vcos/pthreads -I./ -I../libs/ilclient -I../libs/vgfont
-g -c world.c -o world.o -Wno-deprecated-declarations
cc -o hello_world.bin -Wl,--whole-archive world.o -L/opt/vc/lib/
-lGLESv2 -lEGL -lopenmaxil -lbcm_host -lvcos -lvchiq_arm -L../
libs/ilclient -L../libs/vgfont -Wl,--no-whole-archive -rdynamic
rm world.o
```

In fact, although only three commands are being executed there, marked with circles above, if you look closer you'll see that all it's doing is

- Using the C compiler 'cc' to compile the code in hello_world.c into an object file hello.o (look for the text '-o world.o').
- Then using 'cc' to link world.o together with the required libraries to make the final program hello_world.bin.
- And finally using 'rm' to delete the file world.o because it's no longer needed.

The first line is very long because it includes lots of symbol definitions (which have the form -D<symbol>=<value>) that you can see set up in the parent directory's Makefile.include. These aren't used at all in this program but are useful in some of the other demos.

Our new program executable should be found in a new file called hello_world.bin. To see what it does simply type

```
./hello_world.bin
```

If it prints the expected greeting congratulate yourself for having built a successful C program.

Library

Before you can progress to the other demonstrations you'll need to create a number of helpful functions that several of them use. These are in

```
/opt/vc/src/hello_pi/libs
```

which you should 'cd' to. Because this directory is actually in your current directory's parent there's a shortcut you can use:

```
cd ../libs
```

You'll find two directories here, one 'ilclient' which makes the use of OpenMAX (described in the section about multimedia) slightly less hard; and another, 'vgfont', to make the display of vector graphic fonts a little easier. To build these libraries enter each directory in turn and use the make command. The commands you type to do this could be

```
cd ilclient
make
cd ../vgfont
make
```

The first 'make' creates a library file called 'libilclient.a'. Libraries are very similar to object files except they're divided up internally into separate sections that can be ignored if they aren't used. That way libraries can contain the definition of large numbers of useful functions without necessarily making the program that they contribute to any larger than it needs to be. On Unix-style computers this type of library always ends with '.a' (which stands for 'archive') and starts with the three letters 'lib'.

The second 'make' creates libvgfont.a.

Once you've done that you're ready to proceed.

Audio demo

This demonstration is held in

```
/opt/vc/src/hello_pi/hello_audio
```

Again, to build this application simply 'cd' to this directory and run

```
make
```

You should see output of this form from make as the program compiles:

```
cc -DSTANDALONE -D__STDC_CONSTANT_MACROS -D__STDC_LIMIT_
MACROS -DTARGET_POSIX -D_LINUX -fPIC -DPIC -D_REENTRANT -D_
LARGEFILE64_SOURCE -D_FILE_OFFSET_BITS=64 -U_FORTIFY_SOURCE -W
all -g -DHAVE_LIBOPENMAX=2 -DOMX -DOMX_SKIP64BIT -ftree-
vectorize -pipe -DUSE_EXTERNAL_OMX -DHAVE_LIBBCM_HOST -DUSE_
EXTERNAL_LIBBCM_HOST -DUSE_VCHIQ_ARM -Wno-psabi -I/opt/vc/
include/ -I/opt/vc/include/interface/vcos/pthreads -I./ -I../
libs/ilclient -I../libs/vgfont -g -c audio.c -o audio.o -
Wno-deprecated-declarations
cc -DSTANDALONE -D__STDC_CONSTANT_MACROS -D__STDC_LIMIT_MACROS
-DTARGET_POSIX -D_LINUX -fPIC -DPIC -D_REENTRANT -D_LARGEFILE64_
SOURCE -D_FILE_OFFSET_BITS=64 -U_FORTIFY_SOURCE -Wall -g
-DHAVE_LIBOPENMAX=2 -DOMX -DOMX_SKIP64BIT -ftree-vectorize -
pipe -DUSE_EXTERNAL_OMX -DHAVE_LIBBCM_HOST -DUSE_EXTERNAL_
LIBBCM_HOST -DUSE_VCHIQ_ARM -Wno-psabi -I/opt/vc/include/ -I/
opt/vc/include/interface/vcos/pthreads -I./ -I../libs/ilclient
-I../libs/vgfont -g -c sinewave.c -o sinewave.o -Wno-deprecated-
declarations
cc -o hello_audio.bin -Wl,--whole-archive audio.o sinewave.o
-lilclient -L/opt/vc/lib/ -lGLESv2 -lEGL -lopenmaxil -lbcm_host
-lvcos -lvchiq_arm -L../libs/ilclient -L../libs/vgfont -Wl,--no-
whole-archive -rdynamic
rm sinewave.o audio.o
```

The commands have been highlighted again. They

- Use 'cc' to compile the code in audio.c into an object file audio.o.
- Then do the same to compile sinewave.c into sinewave.o.
- Then use 'cc' to link audio.o and sinewave.o together to make the final program hello_audio.bin.
- And finally, delete the files audio.o and sinewave.o.

A new file hello_audio.bin will be created. The code in sinewave creates a table of integers from which the trigonometric sine function can be derived. This is used by the code in audio.c which uses it to produce the digital samples of a note rising and falling in pitch.

The program reads the first argument on the command line as a number which can be zero or one. If the argument is zero audio output is sent to the analogue output (the 3.5mm audio jack plug), otherwise it sends it to the digital HDMI output (which will produce the sound on your display or television).

You can now run this to see what it does:

```
./hello_audio.bin 1
```

Video demo
The video demonstration code is held in

```
/opt/vc/src/hello_pi/hello_video
```

If you look at the files here you'll see the usual Makefile, the C program 'video.c', and a test video file in H.264 format (you'll find something about the different video formats in the section about the XBMC media player). There's also a README file which is required to go with the video file for legal reasons.

Build the program here using 'make' as usual. It'll create a new program called hello_video.bin. You may be able to see these two parts of the command that creates hello_video.bin:

```
-L../libs/ilclient
```

This tells the compiler to look in the directory where we built the ilclient library for any libraries it requires.

```
-lilclient
```

This tells the compiler to use a library 'ilclient'. It will add an initial 'lib' and a '.a' suffix when searching through directories where it expects libraries to be found (that is, it'll look for a file called libilclient.a).

This program also reads the first item on its command line, expecting it to be the name of a file containing video information. So to give it the test file use the command

```
./hello_video.bin test.h264
```

You should see the first few frames of the video 'Big Buck Bunny' appear on your display with no sound.

This program demonstrates the use of the OpenMAX IL interface when used for displaying video. The library we built above 'libilclient.a' is used through the C header definitions held in 'ilclient.h'.

In the section about the XBMC media player we describe the difference between video content and the containing file type. Note that this program plays a raw H.264 bitstream. It won't play encapsulated files (such as MP4 or MKV files).

While the details of setting up OpenMAX IL components and feeding them with data are the province of larger books than this, the function video_decode_test() could be used or adapted in your own code for displaying video.

Texture demo
This demonstration code is held in

```
/opt/vc/src/hello_pi/hello_triangle
```

It demonstrates the use of OpenGL in creating a three-dimensional object (a cube) with textures made from three different images on a display managed by EGL.

The files Gaudi_128_128.raw, Lucca_128_128.raw and Djenne_128_128.raw are miniature (128 pixels by 128 pixels) images of famous pieces of architecture. Each is encoded using three bytes for each pixel of the image indicating the colour of each pixel by the quantity of red in it in the first byte, the quantity of green in the next and the quantity of blue in the last (a raw 'RGB' encoding). These files are read by the program and used to define the 'texture' of the sides of the cube (the same picture being used on opposite sides of the cube).

The tables in the header cube_texture_and_coords.h define the shape of the cube and the main program is in triangle.c.

To see what the program does type

```
./hello_triangle.bin
```

You should see that, having set up the cube, the program then rotates it in three dimensions forever (until the program is stopped, perhaps by pressing the keys 'Ctrl' and 'C' at the same time).

Shader demo
This demonstration code is held in

```
/opt/vc/src/hello_pi/hello_triangle2
```

It demonstrates the use of the OpenGL ES shader language in creating an interactive simultaneous display of the well-known Mandelbrot and Julia sets using transparency.

To see what the program does type

```
./hello_triangle2.bin
```

You should see the picture of a Mandelbrot set in red in the background and a Julia set in green in the foreground. Julia sets are derived from one point in a Mandelbrot set and that point can be changed by moving the mouse. The Mandelbrot set moves to keep the point in question in the centre of the screen and the Julia set changes to display the one associated to that point. The program ends when a mouse button is clicked.

You may be surprised to see two strings in this program that look very much as if they contain programs in some programming language (which is what they are). These two programs are given to the VideoCore processors which compile them and execute them as part of the rendering process. The programming language is the Open GL ES 2 shader language which you can find specified at

```
http://www.khronos.org/files/opengles_shading_
language.pdf
```

The idea behind a 'shader' when applied to a three-dimensional model of a scene built up by OpenGL ES is a little more extensive than simply modifying the colour/shade of parts of the model. In fact these shader programs can perform fairly arbitrary manipulations of both the mesh of points (vertices) defining the 3D surfaces or the area fragments defined between the vertices. On Raspberry Pi they're executed on the VideoCore side of the architecture, not on the ARM.

Because these programs can operate in parallel on many vertices or fragments at once they represent one of the highest performance aspects of Raspberry Pi and in principle could be used to carry out non-graphics orientated tasks, giving access to the VideoCore hardware in a way that's otherwise not possible.

Overlay demo
This demonstration code is held in

```
/opt/vc/src/hello_pi/hello_dispmanx
```

and includes only one C file called 'dispmanx.c'. It demonstrates the use of a Broadcom-specific interface called 'dispmanx' to display a separate transparent overlay on top of the existing display.

To see what the program does make sure there's something interesting in the middle of the display (for example, you could put a window there in the desktop environment), then type

```
./hello_dispmanx.bin
```

You should see that three concentric squares appear for ten seconds: blue inside green inside red. Being transparent you'll be able to move objects (eg windows) beneath them to observe the change in colour they cause.

The program in dispmanx.c that performs this behaviour is very straightforward, with only the main function and no loops or conditionals. It consists mostly of creating the relevant squares in a 200x200 pixel bitmapped image of the screen, which it adds to the centre of the screen before updating it. At the end of a sleep for ten seconds it then removes the squares from the display, updates it and deletes the resources it's been using.

Encoder demo
This demonstration code is held in

```
/opt/vc/src/hello_pi/encode
```

It demonstrates the use of the VideoCore to take a simple uncompressed video format (the program generates an animated test pattern in 'YUV420' format) and encode it into a H.264 bitstream (a task that normally requires a significant computational resource).

To see what the program does type

```
./hello_encode.bin <file>
```

The program ends when a mouse button is clicked. The file that's generated is suitable for playing with the hello_video demonstration program.

Vector graphics demo
This demonstration code is held in

```
/opt/vc/src/hello_pi/tiger
```

where you'll find a number of files, two of them (license.txt and readme.txt) regarding the licensing terms for the program in tiger.c. This program contains an encoding for the vector paths and fill colours that, when drawn, will produce the picture of a tiger's head.

To see what the program does type

```
./hello_tiger.bin <file>
```

You should see a rotating vector graphics image of a tiger's head. The program ends when a mouse button is clicked.

The program in main.c contains an interpreter for the data provided by tiger.c and uses it to generate a long list of Path Specifications (an array of PathData structures) in the function PS_construct. When supplied to the routine PS_render that function uses the correct OpenVG functions to draw each path.

The main function constructs the path specifications and then stays in a loop continuously calling the 'render' function which uses PS_render to draw the tiger on to an EGL surface after setting up a rotation as specified by the variable rotateN, which is updated every time around the loop.

Font demo
This program is held in

```
/opt/vc/src/hello_pi/font
```

You'll find a standard TrueType font, Vera.ttf, in this directory, as well as the code in main.c and the usual Makefile which you can use to build the executable with the command 'make'.

To see what the program does type

```
./hello_font.bin
```

You should see a quickly growing subtitle line of text rendered in the 'Vera' font at the foot of the screen overlaying whatever else you may have there. The text should say 'The quick brown fox jumps over the lazy dog' and repeatedly grow from a small to a large font forever (until the program is stopped, perhaps by pressing the keys 'Ctrl' and 'C' at the same time).

The code uses the font interface provided by the library that was compiled above and isn't very long. After creating a graphics window the same size as the screen it sits in an infinite loop incrementing the 'text_size' variable from 10 up to 50, when it sets itself again to 10.

Each time around the loop the window is cleared to a transparent value (the fourth number in GRAPHICS_RGBA32 is the 'alpha' or transparency value of the colour) and then an area at the bottom of the screen is drawn in green while the area above is drawn in blue. Finally the text is drawn at the right place on the screen to ensure that the bottom of it is at the foot of the screen.

ANNEXES
Index